# JOURNEY INTO CLIMATE

Michael Morrison

# JOURNEY INTO CLIMATE

*Exploration, Adventure, and the Unmasking of Human Innocence*

Paul Andrew Mayewski and Michael Cope Morrison

A DAUNTLESS PRESS

ADauntlessPress.com

Printed in Singapore

*Frontispiece: Nun Kun, Ladakh, Indian Himalayas, 1980. After ascending three icefalls, we crested onto an ice plateau at well over 18,000 feet punctuated by mountains rising yet another 3,000 feet. The journey into this site took many weeks and was filled with intrigue and adventure. More adventure—and a new way to understand climate—was still to come.* (PAM)

*Front and back covers: Views from the top of the Nun Kun Plateau, Ladakh, Himalayas, looking south into the Indian Himalayas.* (PAM)

*This book is available as an eBook from Apple's iBookstore (ISBN: 978-0-9836302-0-3); as a 9 inch x 10 inch paperback (ISBN: 978-0-9836302-1-0); and in PDF format on a CD (ISBN: 978-0-9836302-2-7).*

*A fine art portfolio of 20 images from this book is available. Any image from this book can also be ordered individually at the size of your choice. Please visit OpsinImaging.com for details.*

CONTENTS

*To*

Lyn Weisman Mayewski and Ninon Argoud Morrison

*And to our parents*

Pawel Mayewski and Lisa Asnis

Walter and Sarah Morrison

## PREFACE

THE TWENTIETH CENTURY ushered in "golden ages" of technology, culture, and exploration. It also gave rise to a golden age of climate research, starting in 1957, which revolutionized our primitive understanding of the large-scale processes governing our weather, the ice ages, glaciers, ocean currents, sea level, and the composition of our atmosphere.

The history of climate recorded in snow and ice has played a central role in the astounding discoveries of this period. We, the authors, have been lucky enough to participate in these discoveries, not the least of which was that the snow and ice we researched has trapped within it a remarkably detailed, robust, and accurate record of past climate.

These stories trace the authors' journey through this period and beyond. They begin in a metaphorical, academic tower where we thought all processes were gradual and human activities insignificant, and bring us to today, where we now know climate is moody and volatile—and that human endeavors are currently the dominant force in changing climate.

Beyond the scientific insights and epiphanies, this journey has been intensely personal. At the beginning, we thought of ourselves as curious onlookers, as flies on the wall, satisfying our quest to explore and know more. Today, the mask of our innocence has fallen. We know that since at least the early 1800s, human activities have been affecting climate and that now we are the primary, causative actors controlling it. Today, we know that our individual and collective decisions and actions will determine the evolution of Earth's climate for hundreds to thousands of years and more to come—and consequently the future of Civilization.

We know that humans have an immense impact on climate and that we are responsible to all future generations: The only question is the direction we will choose. We, the authors, have been deeply influenced by this knowledge, leading us to be more activist than has been historically common for scientists. With this book we hope to bring to life the adventures of climate research as seen through our own experiences, the moments when discoveries were made, and a sense of the magnitude of human control of climate.

## HOW THIS BOOK WAS WRITTEN

MOST OF THESE STORIES are in Paul's voice. *Earth-Scale Science*, *Lush and Fragrant*, and *A Volcano on the Andean Spine* are in Michael's. Though individual stories have origins with one or the other of us, we jointly edited all of them, striving for a synthesis of our experiences and vision for the book. It has been an immensely enjoyable experience. We are also grateful to Lyn Mayewski and Ninon Morrison for their many valuable comments and thoughts that have vastly improved the book.

Most of the images in this book are from Paul's collection of slides. Michael scanned and restored these images and prepared all the others, including modern digital images in a variety of formats from low resolution jpgs to large raw files.

Michael also did the layout and coordinated the publication in electronic and paper formats.

Michael Motley, of Santa Fe, New Mexico, provided valuable guidance and review of the design. His knowledge of book production, unerring eye for style and beauty, and typographic expertise were both invaluable and a joy.

Margaret Nagle of Orono, Maine, provided superb copyediting. Karen Powelson of Santa Fe, New Mexico, provided effective and efficient communications with CS Graphics in Singapore. Paul Ignasinski and Ingrooves provided fixed-layout EPUB conversion and electronic distribution.

## ABOUT THE TITLE

AS WE BEGAN WORK ON THIS BOOK, we settled on a title, *Journey Into Climate*, which we thought well reflected our experience and the feeling we hoped to convey. We subsequently realized that we both had ancestors who had also written books with "Journey" in the title. Paul's father, Pawel, wrote, *The Journey and the Pity*, and Michael's granduncle, Walter Nordhoff, wrote, *The Journey of the Flame*. We are intrigued by our continuation of the "journey" theme, present in the writing of our ancestors, and by the evolving nature of the journeys portrayed in each of these books.

# ACKNOWLEDGEMENTS

PAUL ANDREW MAYEWSKI ■ When I talk about our expeditions, I stress the word "our" because all of the more the 50 expeditions into the remotest parts of the world that I led have depended on a strong team for planning, coordinating, preparing and analyzing samples, and interpreting the results.

Our research has been supported by grants from the National Science Foundation, National Oceanographic and Atmospheric Administration, Environmental Protection Agency, Electric Power Research Institute, W.M. Keck Foundation, and gifts and endowments from several private individuals, including Iola Hubbard,, Dan and Betty Churchill, and Brenda Garrand. I have been very fortunate to have teams with extremely talented and impressive people without whom it would have been impossible for us to collect the global array of ice cores that we now utilize for our climate reconstructions. The teams have been comprised of many varied groups: notably, graduate and postdoctoral students I advised while at the University of New Hampshire and now at the University of Maine, more than 40 thus far, including the most recent—Dan Dixon, Bjorn Grigholm, Elena Korotkikh, Susan Kaspari, Erich Osterberg, Mario Potocki, and Nicky Spaulding: many undergraduate students, most notably, James James, Peter Jeschke, Peter Axelson, and Jim Sevigny who accompanied me on the early trips into the Himalayas; remarkable colleagues, such as Tony Gow, Andrei Kurbatov, Michel Legrand, Berry Lyons, Kirk Maasch, Dave Meeker, and Michael Morrison who worked with me to unravel the intricacies and interactions of the diverse array of disciplines necessary to produce, interpret, and disseminate climate change information. Collaborators include Naseer Ahmad from India; Sharad Adhikary from Nepal; Qin Dahe; Shichang Kang; Ren Jiawen; Shugui Hou and Xiao Cunde from China; Ian Goodwin from Australia; Nancy Bertler and Uwe Morgenstern from New Zealand; Jefferson Simoes from Brazil; and Gino Casassa, Ricardo Jana, Stefan Kraus, and Jose Retamales from Chile. Super-talented analytical, technical, and organizational support have been provided by Betty Lee, Sharon Sneed, Cap Introne, Mike Handley, Mark Wumkes, and Mike Waskiewicz. I have also had superb mentors, notably Parker Calkin, Dennis Hodge, Richard Goldthwait, Colin Bull, George Denton, Harold Borns, and Claude Lorius. Amazing support and advice has come from program managers Julie Palais and David Verardo, and from logistics suppliers: the 109th Air National Guard, PICO/ICDS, arrieros (cowboys) in the Andes, Sherpas in Nepal, and porters from places like Tongul in the Ladakh Himalayas, who are no doubt now long gone because the average life span for the men of this village was 35 and 23 for women.

Last and most importantly, I owe a great deal to my parents, and I am truly fortunate to have my wife with whom to share my life, our little family, and plenty of laughter.

MICHAEL COPE MORRISON ■ My father took me on my first climb of a big mountain, Giant Mountain in the Adirondacks. I attribute my love of being outdoors to this, and to the many other outdoor activities including skiing, camping, fishing (though I was never hooked), canoeing, and more, that my mother and father provided for my brothers and me.

My grandfather, Laurence Morrison, introduced me to the wonders of math and science, and especially Earth science. I remember in the early 1960s, when plate tectonics was still a new and thinly embraced idea, walking with him as we watched a storm-swollen stream create small fluvial beds. He remarked on the erosional processes occurring before our eyes, and noted that the beauty of science was that discovery was available to each and all of us by simple observation and experiment. This same sentiment was echoed by my grandmother, Margaret Morrison, who as a Quaker believed that God speaks to each and all of us, and not just to a special few. Margaret and Laurence both held the strong conviction that humanity played a large and central role in the evolution of the conditions on Earth's surface—a prescient belief and one instilled in me at an early age.

I had the good fortune to learn mountaineering from James Goodwin who, born in 1910, had climbed with many early mountaineers, including Fritz Weissner and John Case, and who, at an earlier turn, had also taught my father to climb. My first glacier experiences were in the Canadian Rockies where I climbed with Bill Putnam for two seasons.

Mark Hines and Paul Mayewski guided my way into the emerging discipline of Earth System Science at the University of New Hampshire, and Julie Palais offered me the opportunity to go to Antarctica. Berry Lyons introduced me to the larger history of Earth's climate, and the use of biogeochemical tracers and isotopes to gain both broad and subtle insights.

Paul Mayewski asked me to take on the scientific coordination of the remarkable GISP2 ice coring effort, which plunged me deep into the frontiers of climate science. His faith in me, and support of my work for GISP2, is much appreciated and set the stage for effectively meeting our challenges in bringing GISP2 to fruition.

My parents, Sarah and Walter, have provided inspiration, faith, and unstinting and loving support and encouragement in my education and many endeavors.

*Paul Andrew Mayewski*

*Michael Cope Morrison*

## ABOUT THE AUTHORS

DR. PAUL ANDREW MAYEWSKI is director and professor of the Climate Change Institute at the University of Maine. He is an internationally acclaimed scientist and explorer, and leader of more than 50 expeditions to some of the remotest reaches of the planet, including: Antarctica, the Arctic, the Himalayas, Tibetan Plateau, and the Andes. His scientific achievements appear in more than 300 peer-reviewed scientific publications plus a climate change book with co-author Frank White, "*The Ice Chronicles*", published in 2002. His honors are numerous, including: the inaugural Medal for Excellence in Antarctic Research, awarded by the Scientific Commission for Antarctic Research; the Explorers Club Lowell Thomas Medal; the International Glaciological Society Seligman Crystal; an honorary doctorate from the University of Stockholm; and a private audience in the Forbidden City, Beijing, China. He has developed and led several highly prominent national and international research programs, such as the Greenland Ice Sheet Project Two (GISP2), which included 25 U.S. institutions and during which he hired Michael Morrison as coordinator for the Scientific Management Office; the International Trans Antarctic Scientific Expedition (ITASE), which now includes 21 countries; and numerous outreach efforts, including those with the American Museum of Natural History and the Boston Museum of Science. He has served as either the chair or a member of many national and international research committees, and has appeared hundreds of times in public, including on NOVA, BBC, NPR, several segments with CBS "60 Minutes", as well as many scientific and public lectures throughout the world. He and his wife Lyn, an artist, live in coastal Maine where they enjoy hiking, kayaking, sailing, paddle boarding, walks with their dogs and cats, and planning the next expedition. By 2009, it had been almost 15 years since he worked with Michael Morrison, but meeting at a lecture in Portland, Maine, led to the idea to collaborate on this book and once again use their complementary skills.

MICHAEL COPE MORRISON currently divides his time between his fine art imaging and printing business, and participating in scientific expeditions. He was the first graduate student in the University of New Hampshire's Earth System Science degree program, and was the scientific coordinator for the Greenland Ice Sheet Project Two (GISP2). With his broad interests, Michael went on to study anatomy and bodywork, offering courses and private sessions. More recently, he has offered fine art digital imaging and printing services through his business, Opsin Imaging. He has also trained with The Climate Project and gives talks on climate. fter reconnecting with Paul Mayewski in 2009, he took primary responsibility for production of this book. He has also become an occasional team member for ongoing scientific expeditions. He lives in New Mexico, hiking and enjoying its open ground and big skies.

CHAPTER ONE

# EARLY YEARS

*A ski-equipped C-130 Hercules flies over the Transantarctic Mountains en route to McMurdo from the South Pole.* (MCM)

## REALIZING A DREAM

MY NAME IS PAUL ANDREW MAYEWSKI and I emigrated as a young child with my parents to America on board the *RMS Queen Elizabeth* from my birthplace in Scotland via England. Perhaps the desire to experience remote landscapes started with the early days when I strolled with my parents around the hills of Edinburgh or as we looked out over the vast North Atlantic on the crossing. Growing up in New York City, I spent countless days in museums researching the early expeditions to Antarctica, reading every book and magazine I could find on that continent, and preparing as best I could for an "expedition to the remotest part of the world" that I was sure one day I would make. I dreamed of seeing places that no human had ever seen before and, with luck, of making a discovery. So as a first-year graduate student, with the Sun just rising over the clouds on a sultry New Zealand spring morning in October 1968, I stepped aboard a U.S. Navy Hercules ski-equipped C-130 aircraft in Christchurch bound for Antarctica, and my dreams became a reality. In fact, reality surpassed my dreams because I was to spend many years in total in the field in some of Earth's remotest places.

As the aircraft approached the Antarctic continent, I saw the wispy trails of semifrozen sea ice and the vast Transantarctic Mountains where I would spend many months traveling and camping that austral summer. I was deeply aware that I left a world in turmoil, one in the midst of Vietnam and the Cold Wars, fraught with tremendous personal and circumstantial challenges for so many. This awareness severely tempered my excitement and, as I watched this new landscape unfold before me, I felt rising in me a profound resolve to make a contribution—a resolve to justify the opportunity and good fortune that had brought me to this day.

This resolve took up permanent residence in me. It has carried me through many difficult challenges and has compelled me to always look deeper and to push farther.

On that day we flew over Northern Victoria Land where, as yet unknown to me, I would lead one of the earliest expeditions into this vast and extreme landscape. While Northern Victoria Land waited, on this first expedition I would gain experience that would allow me to mature into an expedition leader and a scientist, living in the vastness, the quiet, and the raw exposed beauty of Antarctica.

*View from a ski-equipped C-130 aircraft of sea ice melting along the coast of Antarctica with the Transantarctic Mountains in the distance. (PAM)*

*View from a ridge in the Transantarctic Mountains looking into the "ice-free" Dry Valleys. Here the mountains form a barrier against the advance of the East Antarctic Ice Sheet.* (PAM)

*One of our camps in the Transantarctic Mountains. We were four people with two snowmobiles and four Nansen sleds. Supplies were carefully spread among the sleds so that, if one were lost, we would not lose critical items.* (PAM)

*Right: Traversing a snow slope where the Antarctic Ice Sheet (left) meets a snow apron draped off the Transantarctic Mountains (right). The area is characterized by strong winds and severe drying due to the extremely low humidity. Mountain barriers like this one can force deep ice that is pushing coastward from inland sources toward the surface. Consequently, very old ice is exposed (a million years or more in some places) and embedded meteorites, Antarctica's extraterrestrial treasure trove, are revealed.* (PAM)

## VAST AND MERCILESS

WITHOUT MY USUAL GAUGING TOOLS—trees, people, houses, bridges, I found it hard to comprehend the vastness of Antarctica. But there I sat, at what most of the planet's population thinks of as the bottom of the world, so far from anything we as a species have experienced or even cared about, struggling for a sense of scale. I found myself reviewing what I knew about Antarctica's dimensions: one and a half times the size of the United States; a massive, bisecting mountain range dividing what is called East Antarctica from West; 70 percent of the Earth's fresh water; and 90 percent of the ice—enough to raise sea level well over 200 feet if it all melted. In contrast to the Arctic, which is an ocean surrounded by the North American and Eurasian Continents, the Antarctic is a continent entirely surrounded by the Southern Ocean. In the Arctic, a thin skin of sea ice, rarely more than several feet thick, is formed by the frigid winter temperatures. The Antarctic land mass provides a solid platform on which snow falls and never melts (except near the coast) to form the Antarctic Ice Sheet, as much as 15,000 feet thick in places, which flows in slow-motion rivers of ice from the interior to the surrounding sea.

The hundreds and thousands of icebergs below me, each of which I knew to be the size of city blocks, appeared as so much confetti. I traced their origins in my imagination, beginning with snowfall that piles higher and higher and eventually tops the high passes of the Transantarctic Mountains, then flows to the Ross Sea Embayment where it meets other ice streams to form the Ross Ice Shelf, a 1,000-foot-thick slab of eye-burning white ice floating in the Ross Sea. This slab, pushed northward by the steady drive of new ice from the plateau, sloughs off great chunks at the Shelf's northern edge—the icebergs I now saw. This "calving front" forms a near vertical face extending hundreds of feet above—and many more hundreds of feet below—the sea surface, and was descriptively called "The Great Barrier" by early explorers. With this grand flow of ice in mind, I also thought of Captain Scott and his expedition mates who died in 1912 on this ice shelf and, like the ice itself, eventually plummeted at the barrier into the sea.

*Looking down the Beardmore Glacier, a massive outlet glacier that drains ice from the polar plateau to the Ross Ice Shelf. Mountain relief in the distance is thousands of feet.* (PAM)

*Left: Icebergs, the size of city blocks and larger, calving (discharging) from the edge of the vast Ross Ice Shelf (mid-picture) with Mt. Erebus in the background, the continent's only active volcano.* (PAM)

Even with these facts and stories in mind, I still had difficulty grasping the scale of what I saw out the window, like the many others who have described the same perceptual challenge. The experience of this vastness has continued throughout my many years of exploration on foot, by ski, snowmobile, tractor, and aircraft. It is one of my most enduring impressions of Antarctica: mountains that seem close enough to touch but which require many days of travel to reach, and what—from a distance—appear as small hills turning out to be giant cliffs of several thousand feet and more.

Thinking of many of the early Antarctic expeditions of the late 19th and early 20th centuries, I reminded myself that Antarctica is as merciless as it is vast, and promised myself to maintain an appropriate humility. This expedition was the beginning of many opportunities to experience the whims of Antarctica in the form of gale-force winds, bone-chilling cold, and hidden crevasses, interspersed with amazing vistas, solitude, and peace. A full respect of, and love for, this wild, immense continent was to seep deep into my bones and be regularly renewed on each visit.

*Left: The Commonwealth Glacier flows onto the floor of one of the Dry Valleys, Taylor Valley, in coastal East Antarctica. Commonwealth is fed by snow accumulation in the nearby mountains rather than through ice flow from the vast interior ice of the Antarctic Ice Sheet.* (PAM)

*Below: November 1968. Boulders deposited by previous, more extensive glaciers have lain exposed at the surface for tens to hundreds of thousands of years, resulting in whimsically wind-carved shapes. Paul Andrew Mayewski on his first expedition to Antarctica. Photo by Parker E. Calkin, field leader and Emeritus Professor, University at Buffalo.*

## WE THOUGHT ANTARCTICA WAS TIMELESS

ON MY EARLY EXPEDITIONS TO ANTARCTICA, I set out to study the slow, gradual processes that—as we believed at the time—etched the planet we now see. I went to find the mechanisms by which mountains and ice sheets are built; to understand why slopes have their particular shapes; and to learn how the sensual curves and dramatic skylines of our stunning world came into being

Evidence for the accepted wisdom of imperceptibly slow changes was all around me in the Antarctic landscape. Rocks that had remained in one place so long that wind-carried dust had sculpted them into bizarre and beautiful shapes; mummified seal carcasses 1,000 years or more old; and the ice sheet itself, estimated to have been present for close to 30 million years. In this place, where temperatures reached -130 °F, with almost no moisture in the air, intense winds, and extreme remoteness, surely "gradualism" must reign.

However, on my second expedition to the Transantarctic Mountains two years later, I found evidence that coastal ice was retreating, perhaps as recently as the last few hundred years and at a rate much faster than we had imagined possible. At the time, I didn't realize that finding processes that occurred faster than we had imagined would characterize my entire career, as well as the careers of climate scientists around the world. At the time, I didn't know that 30 years later, we would have satellite images of thousands of square miles of ice shelves disintegrating in the span of weeks, with the newly exposed, darker-than-ice water absorbing the Sun's heat, leading to another round of disintegration.

Today, we know Earth's climate for what it is—moody and volatile. It has shifted in the blink of a geologist's eye—and even a human childhood—from cold and windy to hot and still, from building ice sheets miles thick and lowering sea level below the continental shelves to melting them and flooding coastlines. It is perhaps because Earth's climate has been relatively quiescent over the past 10,000 years that, until the late 1980s, humanity and scientists alike thought climate changed only very slowly. We also now realize that Civilization itself is most likely a product of this unusual time that is relatively free of climatic tantrums. But even within this unusually stable period we were to discover there have been "small" climate changes that have had profound impacts on the course of civilizations and ecosystems, testifying to the fragility of our lifestyles.

*A mummified seal in the Dry Valleys of the Transantarctic Mountains. Juvenile seals that become disoriented wander into the valleys and die. They mummify in the very low humidity and strong winds. Many such seals have been exposed like this for hundreds to thousands of years.* (PAM)

*Opposite: Winds and salts combine to etch bizarre, sculptured rocks like this vone that is nearly 10 feet high in the Dry Valleys.* (PAM)

## SCOTT TENTS STILL IN USE

BEFORE HEADING TO THE FIELD ON EVERY EXPEDITION, I would take our team out for a few days to test the snowmobiles, sleds, and other equipment. The huts of the Captain Robert Falcon Scott and Sir Ernest Shackleton expeditions were only 30 miles from McMurdo, the main base of U.S. research operations in Antarctica. They made a convenient destination for these "shakedown" trips. Today, these huts have been preserved, but on my early trips there, they were partially filled with snow as if Scott or Shackleton had just left and forgot to latch the door.

Visiting these huts, I thought of how far we had come since Scott, Amundsen, and Shackleton's "Golden Age of Antarctic Exploration." In 1929, the "Little America" stations ushered in an era of major scientific exploration and the use of supporting aircraft. In 1957, The International Geophysical Year (IGY) laid the logistical and scientific foundations for the current era of exploration—and what became the Golden Age of Climate Research.

The nature of an Antarctic expedition had also changed radically. In Scott's day, several years were required just to get to Antarctica and back—when they did come back—and the rock samples collected during his ill-fated expedition were as highly prized as those from the Moon are today. By the late 1960s, a few days of travel by airplane brought me from North America to McMurdo, and I would be back home just a few months later. How dramatically our world and lives had changed in just a few short decades.

In the face of these changes, I was struck by the fact that we were still using designs for tents, stoves, and sleds similar to those used by Scott at the beginning of the century. Antarctica was, and is, such an extreme environment that a mix of past—tried and true—and modern technology is required. Today, Antarctica still demands the spirit of exploration and perspective it did at the beginning of the 1900s. Because Antarctica is so unforgiving and so challenging, there remains much to explore to this day.

*Our snowmobile shakedown trip destination—Captain Robert Falcon Scott's hut at Cape Evans, erected for the 1910–13 British (Terra Nova) Antarctic Expedition.* (PAM)

*Boxes from the 1910-13 British Antarctic Expedition still rest undisturbed in the British Antarctic Expedition Hut.* (PAM)

*Right: Snow inside the 1910–13 British Antarctic Expedition hut as we saw it in 1970. Today, the huts are carefully sealed each winter to preserve the items inside.* (PAM)

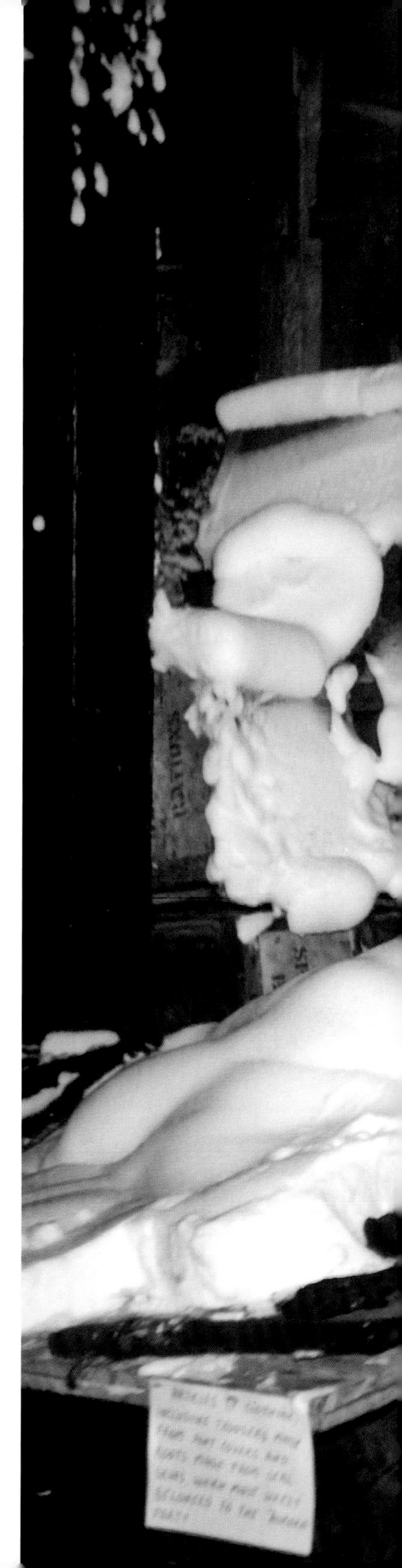

FRY'S
PURE CONCENTRATED
COCOA

## TEA, COOKIES, AND CARBON MONOXIDE

OUR TENTS WERE PERCHED ON THE EDGE of the polar plateau, 9,500 feet above sea level, overlooking the vast Ross Ice Shelf with views that stretched hundreds of miles. Crystal clear air filled my lungs and the wind blew just enough to make me feel invigorated, alive with a sense of place and adventure. This was one of those days when I knew, beyond a doubt, why more than a hundred, isolated days per season in a tent were worthwhile. I was ecstatic to be in Antarctica and a part of its adventure.

The storm that formed that day was no different from others we had experienced and, as the winds hardened and the snow accumulated around us, we thought we would be safe in the Scott tents that had sheltered us—and so many others—so many times before. We settled in for the day, making tea, eating cookies, and talking about the field season and our homes. We were in our early twenties and although experienced with a few Antarctic seasons, we had forgotten a measure of the humility Antarctica demands.

As the storm blew, and the tea and cookies were finished, we felt tired and began to nap. My rest was disturbed by the tea and the subsequent call of nature. As I stepped from our shelter into the storm, I collapsed to the snow from the rush of oxygen. The vent atop the tent had not kept up with our stoves and breathing and we had been falling into a carbon monoxide stupor. We were lucky. Without the tea, we might have slept just long enough to never awaken.

I was humbled by the youthful mistakes we made that day. We ignored the effects of our own actions and poisoned the air we breathed. I never forgot the lessons I learned that day—nor my luck. To this day, that event lives in me, not just in the form of good survival skills, but as an analogy of humanity's effects on our atmosphere—and of our ignorance and youthful confidence.

*Scott tents in drifting snow during a strong wind. (PAM)*

CHAPTER TWO

# A TASTE FOR EXPLORATION

*Checking the ropes that connected our four-person team as we crossed crevasses in Northern Victoria Land.* (PAM)

BY 1973 MY PH.D. WAS COMPLETE. For it I had spent several field seasons working throughout the Transantarctic Mountains collecting data to identify the past extent of the Antarctic Ice Sheet and develop an understanding of how fast modern glaciers in these mountains were moving.

Many of my early results fit beautifully with the then current belief that Antarctica was timeless, or at least unchanging on timescales significant to humans. It appeared to us that the Antarctic Ice Sheet only changed size over tens of thousands to millions of years, and that the smaller mountain glaciers adjacent to the ice sheet were frozen to the rock below them, making any motion very slow.

In June of 1974, while at my post-doctoral position at the University of Maine, I received a letter from the National Science Foundation. It said my proposal to study glaciers in Northern Victoria Land was funded. I was very excited: This northernmost portion of the Transantarctic Mountains was still unexplored, and I hoped that perhaps, in this unknown region, I would make that big discovery.

Beginning in October that year, I led our four-man team for just over three months by foot and by snowmobile as we traveled more than 1,500 miles in this remote and unexplored place. We had two snowmobiles and four Nansen sleds that held our food, tents, fuel, and scientific equipment for the entire journey. Nansen sleds are superbly crafted wooden sleds that bend over irregular snow surfaces and hold several hundred pounds. Easily repaired with slats of wood and twine, they have stood the test of time since their use in the Arctic in the late 19th century by the Norwegian explorer Fridtjof Nansen. As we traversed Northern Victoria Land, we exercised great care to avoid crevasses, prepare for storms, and prevent tent fires, while also observing and absorbing what this new place had to tell us.

On this trip, Antarctic fervor truly set in. I loved living in the environment I was studying, traversing glaciers and scaling unclimbed mountains as part of my job. Perhaps the fervor came from experiencing the solitude of being one of only a few creatures—not just humans, but any creatures—within hundreds of miles. Or perhaps it was the way my head cleared, casting away the complexity of life in Civilization for the simplicity of a tent that sheltered me from the wind, of evening meals comprised of melting ice for water and one-pot dinners, of hearing only the wind, or of nights spent excavating old memories. (Whoever said we never truly ever forget, but rather just cannot easily recall everything was correct, for I could conjure up memories of days in grammar school and childhood scenes, memories forgotten to me at any other time.) Or, perhaps, the fervor was coming from the Antarctic water—wild, cold water that was slowly replacing the "civilized" water in my body.

*Left: The Rennick Glacier in Northern Victoria Land, with our snowmobiles and sleds in the foreground. (PAM)*

*A giant wind-scoop carved into the side of a Transantarctic Mountains glacier. Note person for scale at the top middle of the photo. (PAM)*

## CREVASSES AND STORMS

AS THE ANTARCTIC ICE SHEET FLOWS TOWARD THE OCEAN, it passes over mountains, through valleys, and, at its coastal edges, begins to float. As the ice moves along its path, cracks known as crevasses form at the bends, bulges, and other junctures along the way. They can range from a few inches wide and a few feet deep to 30 or more feet wide, and well over 100 feet deep, and can extend for miles in either direction—capable of consuming anything that falls into them, including large vehicles.

In Northern Victoria Land, we came across many crevasses. Some were obvious and prompted a long detour. There were, however, other crevasses that were not just inconvenient, but downright terrifying. In these, wind-blown snow bridges covered the surface openings, making them all but invisible to us as we traveled—gargantuan vehicle traps waiting for us to make a mistake.

Today, though not completely foolproof, we use satellite imagery and portable ice-penetrating radar to guide us. During our Northern Victoria Land travels, we relied on slow, careful movement and on our senses that became acutely attuned to subtle clues: slight differences in topography, faint shadows, the surrounding terrain, and the intuition that grows day by day and year by year. We had some close calls, but never a serious crevasse incident.

Northern Victoria Land is the meeting point between the cold and dry north-flowing air from the high interior of the Antarctic Ice Sheet; the relatively warmer, wetter, south-flowing air from the bordering Southern Ocean; and the circumpolar wind that divides the two. As these air masses pushed against each other, the boundary between them moved northward and southward. When the Southern Ocean air was gaining dominance, as it did every couple of weeks, we would first notice tiny swirling patches of light snow. These, we quickly learned, were soon followed by hurricane-force winds that could last several days. The warmish, near-freezing snow of these storms penetrated the tiniest open spaces in our clothing and tents, and easily melted, leaving us wet and cold.

We soon came to miss the much colder, drier conditions that came with the interior air masses. Though this air was often -40 °F or colder, we stayed dry—and could more easily warm ourselves. It was during one of the warmer, wet, piercing storms that two of us nearly froze. Our clothes and tents were so soaked we could not dry them with our tiny cooking stoves. The storm lasted three days and left us barely strong enough to travel the 20 miles of crevasse-filled terrain to our dry base camp tents.

During another such storm in 1974, as we huddled in our tent prizing anything dry and warm, we realized that Northern Victoria Land, situated under this push-me-pull-you of air masses, was a transitional area. We reasoned that if we could somehow determine how the proportions of cold-dry air to warm-wet air had changed over time, it might tell us about larger changes in atmospheric circulation for all of Antarctica. Talking this over in our tent, I filed it away as an interesting possibility. I did not know then that it would become a cornerstone of my research, and one of the more powerful strategies we have for understanding climate history.

*A storm rages on the Antarctic ice plateau. Our tents often became half buried. (PAM)*

*Left: Strong windstorms could rage for days, requiring us to carefully secure our tents and gear—and constantly check them. John Wilkinson (then at The Institute of Polar Studies, Ohio State University) illustrates how we secured the tent to ice screw anchors set into the glacier in the midst of a major storm that pounded our tent and came in waves every few minutes that almost sounded like a freight train. (PAM)*

*Right: When we traveled across crevassed areas, we stopped often to evaluate the terrain and our route. (PAM)*

*Aerial view of crevasses produced as ice stretches out from the Antarctic ice plateau to the Ross Ice Shelf. (PAM)*

## FASTER THAN WE THOUGHT

MOST DAYS WE WALKED FOR SEVERAL MILES OR MORE across a landscape of ice, snow, rocks, and dirt left by glaciers, and mountains untouched by anything but wind, snow, and sun for millions of years.

During field seasons from 1968 into the mid-1980s we explored significant portions of the vast, 1,500-mile-long Transantarctic Mountain Range. Some days it was so cold and the wind so strong that I could feel every bone in my body. These mountains act as a barrier, impeding the flow of the plateau ice, over 6,000 feet thick, as it presses toward the Ross Sea, allowing only a few passageways for "outlet glaciers". One of these outlet glaciers, the famous Beardmore Glacier, provides a route to the broad plateau of the East Antarctic Ice Sheet—albeit a crevassed and dangerous one—used by Scott and his team during their 1911–12 traverse to the pole.

On our expeditions, we made many observations to determine the current state of the Antarctic Ice Sheet. One of the key observations we made was the location of surface debris on valley walls or stretched out on flats in front of the modern glacier edges marking the earlier positions of "grounding lines," the place where the bottom of an outlet glacier floats off the rock below as it reaches the sea. When the surface debris that marks the position of a former grounding line is farther inland and lower in elevation than older remnants, it indicates warmer conditions or higher sea levels than in earlier times and vice versa. By observing the past and present locations of these grounding lines, we learned when and how fast the ice sheet had changed size in the past.

Our studies showed a large-scale retreat of outlet glacier grounding lines starting about 10,000 years ago, in agreement with our understanding that the North American Ice Sheet started to melt around 14,000 years ago, raising sea level about 300 feet over the next few thousand years. This rise in sea level would have lifted Antarctic ice shelves and increased the size of the floating outlet glacier tongues, driving Antarctic grounding lines inland. Freed from contact with the ground, ice would have flowed more rapidly toward the sea, and the relatively warm ocean water would, in turn, have melted the floating ice from below. All these factors led to an overall reduction in the size of the grounded area of the Antarctic Ice Sheet.

There was, however, a troubling detail in our observations: It appeared that smaller glaciers and snow patches in the mountains adjacent to the outlet glaciers, and the outlet glaciers, were also changing. They were all retreating. Was the retreat a continuation of the retreat that had been going on gradually for thousands of years, or was something new happening? After all, the small glaciers, as a consequence of their size, were more responsive than the ice sheet and presumably would have retreated first. Small glaciers typically respond faster than larger ones and, if anything, we would have expected them to be advancing today since open ocean was closer now than in cold, glacial times leading to an increase in snowfall on the small mountain glaciers and snow patches.

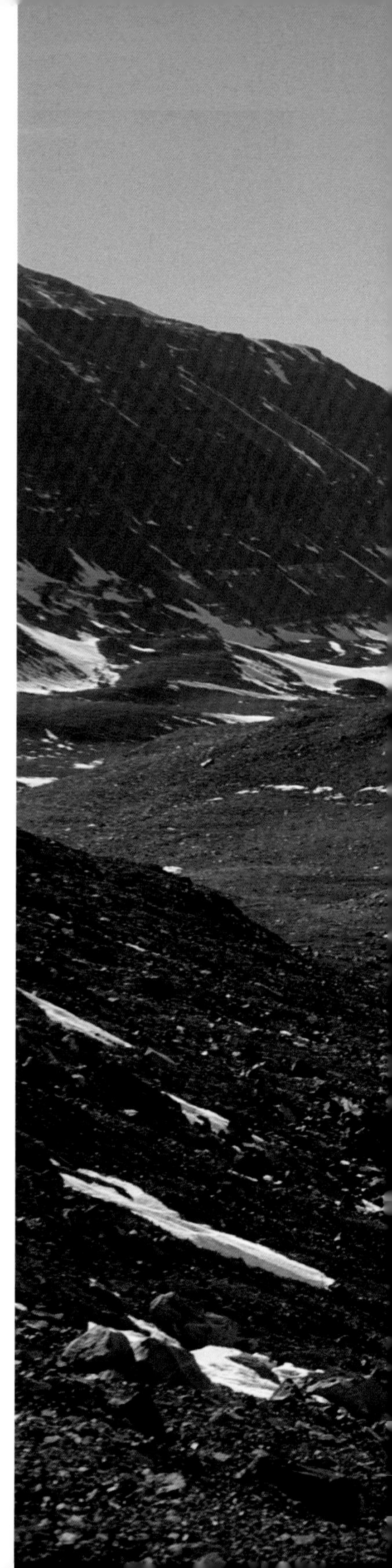

*The edge of a glacier in Northern Victoria Land. The frozen pools at its base are formed from summer melting, helped along by the heat absorbed by the dark rocks in the area.* (PAM)

I remember sitting in my office back home, studying the results we had collected. If the data was to be believed, there was some process happening—in the present—that was accelerating the unexpected retreat of these small glaciers. As I discussed this question with my colleagues, all of us still in the grips of a "gradualist" mind set, we could not make sense of my observations, and the most obvious conclusion was that there was something wrong with my data. But I had been there and seen the retreat with my own eyes. While I didn't know what to make of it, neither did I believe it was wrong.

We published the results as they were and suggested some modern, as-yet hypothetical process as the responsible agent. Today, we know we were seeing the beginnings of human-caused warming of the 20th century. At the time, I hadn't shaken gradualism, but the seeds of doubt had been sown. Who knew into what they would grow.

*Magnificent mountain massifs comprise the Transantarctic Mountains. (PAM)*

## A THREADBARE TENT

FIFTY DAYS INTO OUR 1975–76 EXPEDITION to Northern Victoria Land, one of our Scott tents became threadbare and, I felt, no longer reliable. We pulled out the spare tent and were startled by its bright orange color compared to the old one. I was sure that the faded tent was new at the beginning of the trip and wondered why it had seemingly self-destructed. At the time, I had no answer and wrote it off as defective fabric. We continued our travels with the new tent and thought no further on the matter until the early 1990s when, as I reviewed old slides for a presentation, I found a picture of that faded tent. This time, though, I knew what had happened.

In 1985, the British scientists Joseph Farman, Brian Gardiner, and Jonathan Shanklin found disturbingly low levels of high-altitude ozone over Antarctica based on measurements of ozone made in Antarctica that started in 1957 during the International Geophysical Year (IGY). They also found equally disturbing high levels of ultraviolet light at the Earth's surface—ultraviolet that was normally blocked by the now-missing ozone.

Ultraviolet light, or UV, is extremely destructive. If all the UV that arrives at the top of the Earth's atmosphere were to reach the ground, it would quickly kill most, if not all, plants, animals, and oceanic phytoplankton—phytoplankton that are responsible for a significant amount of the atmosphere's oxygen and form the base of the marine food chain. To say that a loss of the high-altitude ozone that blocks UV was a development of profound urgency is an understatement.

The discovery of the "Antarctic ozone hole" set off—appropriately—a firestorm of study to understand whether this was a "normal" event or something new, what was causing it, how it might progress, and what we might do about it. Meanwhile, atmospheric chemists, including Susan Solomon, Mario Molina, F. Sherwood Rowland, and Paul Crutzen, had shown that the widely used class of industrial molecules known as chlorofluorocarbons (CFCs) was a highly efficient, ozone-destroying catalyst in the high atmosphere over Antarctica and were responsible for the current state of the ozone hole. Because ozone itself is not trapped in snow and ice, a French colleague, Michel Legrand, and I examined compounds that are associated with ozone and are trapped in snow, and thus provided a proxy for historical ozone concentrations. In 1990, as a small contribution to this effort, we used records of this chemical tracer we collected from a 20-foot-deep snowpit, that we sampled continuously down every half inch, located about 25 miles from South Pole. Other records we had collected earlier from several sites in East Antarctica demonstrated that levels of ozone were much lower in recent decades than they had been at any time in the past 10,000 years.

A large portion of Earth's species, and indeed Civilization itself, was spared by the rapid, coordinated, worldwide response to the Antarctic ozone hole crisis—the Montreal Protocols prohibiting the manufacture and use of CFCs. Today, in 2011, the long-lived CFCs from before the ban are still destroying ozone, but their effect is steadily decreasing and UV levels are in decline. Without the swift, global CFC ban, we would now be well on our way to an unimaginable catastrophe.

The ozone hole deeply rattled my sense of gradualism and the ability of humans to affect global processes. It also dramatically altered my belief that Antarctica's distance from the industrialized Northern Hemisphere meant it was safe. Human industrial activity had had a global effect of monumental proportions and, even more significantly, coordinated human response had effectively prevented an emerging catastrophe. It is a tale of two worlds—what might have been, and what is, because of Civilization's effective response.

Our threadbare tent turned out to have been an early result of the then undetected ozone loss. I think of it often, as I do my experience of carbon monoxide poisoning. Ozone loss is not the only significant effect of human activity, and assessing our impacts has become a hallmark of my research—part of the contribution I hoped to make in return for the luck I have had to spend my life exploring Earth's remotest reaches.

*Sergei Abakumov (then a Soviet exchange scientist) and John Attig (then a University of Maine graduate student) having lunch next to the threadbare tent.* (PAM)

375 PR
FLAGS, GLUE,

## A LONG VIEW

AFTER PORING OVER MAPS AND REVIEWING our observations, after long flights to McMurdo and endless lists of items to pack and prepare, after loading the C-130 and flying to the Transantarctic Mountains, in 1980 we were finally traveling up the Rennick Glacier in Northern Victoria Land, to return to the accumulation stakes we had placed there three years earlier.

While our grounding line and small glacier studies gave us a glimpse into the past behavior of these glaciers and ice sheets, we really wanted to know, in much greater detail, what they were doing in the present. We had placed wooden stakes in the glaciers, marked where they intersected the snow surface, and were now returning a few years later to see how much snow had accumulated in the intervening period. As we climbed the Rennick to the first stakes, our hearts sank. We saw two of our stakes lying in a gully, having been snapped off by the strong winds that funnel cold air off the high glacier surfaces. We would get no data from these stakes.

Other stakes had survived, and we stood for hours, trying to stay warm, eyes tearing from the wind, carefully turning small knobs on our surveying instrument to accurately determine their positions, without accidentally bumping the instrument and losing the day's work.

The accumulation stake studies were only one of the many observations we made, and they, in turn, were a part of a much larger monitoring effort, begun earlier in the 1957–58 International Geophysical Year (IGY). The idea was simple: Like a doctor measuring your blood pressure on each visit, make regular observations on an ongoing basis and, after a period of time, a picture of any changes would emerge.

Many monitoring projects began during IGY, including the U.S.-led atmospheric carbon dioxide measurements and the British-led high atmosphere ozone measurements, both from the vantage point of Antarctica. Once IGY ended, acquiring the funding for monitoring was difficult. Still in the grips of gradualist thinking, there was an argument that monitoring would only validate that there were no significant changes occurring and was thus a waste of resources. Because of extraordinary efforts by many, including Charles David Keeling of Scripps Institute of Oceanography who ensured that atmospheric carbon dioxide monitoring continued, we have monitoring records of utmost importance. Keeling's carbon dioxide record showed the yearly cycle of plant growth, and the steady addition of fossil-fuel carbon dioxide to the atmosphere. When he started observations in 1958, carbon dioxide was at 315 parts per million. Today, in 2010, levels have reached 390 parts per million.

By the mid-1970s, I was among an increasing number of scientists advocating for, and writing proposals with a strong emphasis on, monitoring. The emerging awareness that humans were capable of significantly altering local and global conditions had begun to take root, including the initiation of Earth Day in 1970. Monitoring alone, however, does not tell the whole story: We must be able to explain what monitoring tells us and, critically, we must have an historical context.

Superb work was being done to explain the Keeling carbon dioxide record, with its annual cycle of northern hemisphere vegetation growth, overlaid by the steady increase from fossil-fuel combustion. But the question of a longer view, one that would stretch thousands and even hundreds of thousands of years into the past, kept nagging at me and others. As it turned out, since my first step on the ice of Antarctica in 1968, the answer was beneath my feet. Trapped in the ice itself was an exquisite record of past climate. I was to become one of a group of scientists from around the world who would collect cores of ice at promising locations and extract from them histories of climate. These histories, some almost a million years long, would provide just the perspective needed to understand modern observations in their long-term settings.

Ice cores provide truly remarkable and robust records of past temperature, precipitation, chemistry of the atmosphere, sea ice extent, volcanic activity, forest fires, marine and terrestrial biological productivity, and more. Ice cores are cylinders of ice (typically two inches to four inches in diameter, sometimes larger) that are extracted a few feet at a time using a variety of drilling instruments ranging in size from handheld lightweight systems that we carry into the mountains to massive rigs requiring weeks to assemble and significant logistic support. The ice cores are age-dated like giant tree ring records—physical and chemical measurements from the ice cores mark the seasons and years extending back in time for periods from decades to over 100,000 years.

Precise correlations of features identified in our ice cores with known historical records of temperature, volcanoes, droughts, famines, forest fires, and nuclear accidents and explosions have extended our confidence in these ice core histories. To reconstruct climate from these ice cores, we calibrate the measurements with humanly monitored changes in temperature, wind speed, precipitation, sea ice extent, and other climate data covering the last few decades and in some cases as far back as the turn of the 19th century, plus historically, geologically, and anthropologically recorded droughts, famines, volcanic eruptions, forest fires, aboveground nuclear explosions and more. Our ice core records have taken us on an amazing expedition through time and attracted a large number of researchers in other disciplines to join us, including mathematicians, physicists, chemists, biologists, anthropologists, artists, historians, and musicians who share their questions and their talents. Through a single scientific methodology, the physical adventure of finding and recovering ice cores suddenly merged with the adventure of scientific discovery.

*Getting close to the end of a long day of sledging.* (PAM)

CHAPTER THREE

# THE HIMALAYAS

*Punjabi children at a roadside rest stop on our way into the Himalayas.* (PAM)

## WHERE THE PEOPLE LIVE

BY 1977, I HAD BEEN WORKING IN ANTARCTICA FOR NEARLY A DECADE—a decade in which my desire to explore had steadily grown, fueled by the tales we told each other in our tents waiting out storms, by the satisfied exhaustion of long treks through the mountains, and by the continued gnawing drive that there was something critically important waiting—needing—to be found. After a U.S. senator visiting Antarctica made it very clear to me that our research could spawn significant applied value, I was driven to identify the best prospects for making use of my desire to produce science that justified public trust and funding.

My attention began to turn to Asia, where in contrast to the unpopulated Antarctic, nearly half the world's population lives. I thought that perhaps extending my research to places where people live would be a logical step toward conducting science that might help people. As I was to find out in later years, our expeditions into even the remotest parts of Antarctica had results that were of great value to humanity and Earth's ecosystems, but this realization was to require understanding climate at hemispheric and global scales.

The Asian mountain system, like Northern Victoria Land, is at the confluence of several air masses: polar air from the north; maritime air from the Indian and Pacific oceans; and continental air from central Asia. The monsoon, controlling so much of daily life in the region and supplying nearly all the water, is a result of the back and forth between these air masses. Thinking of what I had learned in Northern Victoria Land, I wondered what an ice core history might tell us about the push and pull of these air masses—and what it might tell us that would be valuable to the billions of people living there. Although there were some temperature and precipitation records from countries like India going back to the mid-1800s, they were mostly from the lowlands and, like all such records, were limited in type and scope. We hoped to produce a detailed and long history of climate for the whole region.

Over the next few years, we made many expeditions to the Himalayas, gaining insights into the transport of acid rain pollution around the globe, as well as the monsoon and its control on the water resources of Asia. It was, however, not easy work. On my first Asian expedition in 1979, I set out with one undergraduate student, Peter Jeschke from the University of New Hampshire, where I was then a professor, to recover the first ice cores from the Indian Himalayas. We each lost nearly 30 pounds in a few weeks—the combined effect of elevation, exertion, and local food. While not so remote as Antarctica, we did not have the possibility of support from a large scientific base and aircraft, and we carried all our camping gear, scientific equipment, and food on our backs.

One insight among the many that resulted from these early expeditions was the effect of global warming on the glaciers themselves. During these early years, we were able to find ice containing a high-quality record—ice that did not experience any significant melting—at elevations as low as 15,000 feet. Today, in 2011, we often have to climb close to or above 20,000 feet to find such glaciers. Well-preserved ice core records at lower elevations are now gone—victims of increased heating of Earth's surface caused by rising levels of carbon dioxide from fossil-fuel combustion.

*A lunch stop in the Punjab as we traveled north into Kashmir and the Himalayas. (PAM)*

*Locals cleaning clothes in Dal Lake, Srinagar, Kashmir, seen during our transit into the Himalayas.* (PAM)

G. A. BUDOO
MANUFACTURER & DEALER OF
MBROIDERIES, SHAWLS, SILKS
SAREES & NAMDAS ETC.

THE VERY NEXT YEAR, 1980, I returned to the Himalayas with a team of University of New Hampshire students, colleagues, and researchers from Aligarh Muslim University. I had been preparing for this expedition to Ladakh for several years, arranging funding and permission to enter the region. In looking for information on the area, I read Fanny Bullock Workman's book detailing her 1906 ascent of the Nun Kun massif in Ladakh, India, close to the borders with Pakistan and Tibet. From her descriptions I was able to identify a promising site for an ice core. There, on the Nun Kun Plateau, we would step up our effort to understand the history of the Asian monsoon. We would literally walk in her footsteps on this expedition and collect a history of climate over the time separating our visits—and before.

The Nun Kun site was high, over 18,000 feet, and like Northern Victoria Land, at the boundary of several air masses. In spite of the months and years spent identifying the location, I was still stunned when we reached the Zoji La Pass. To the south was the endless expanse of forests and green we had just traveled through, and to the north was the arid and dusty landscape of Ladakh stretching as far as we could see. The boundary we sought to study was laid out before us in bold relief.

Having identified a promising site, our work, it seemed, had just begun. Well over four weeks of bus and truck rides, border crossings negotiated by

required bribes, and adventures buying equipment, fuel, and food, brought us to the remote village of Tongul—the last human outpost before the final ascent to Nun Kun that would take us yet another three weeks to climb.

Tongul was cut off from the world much of the year by winter snow and villagers were extremely wary of foreigners. We were amazed by the work they had begun on a road, breaking large rocks into gravel by hand. At that time, they were extending the road by only a few meters per year. As best we could tell, the average life expectancy for women in the region was mid- to late 20s; I met a man my age who looked far closer to 70 than early 30s.

We were counting on being able to hire some villagers to help us carry our loads to the mountains and hoping that they might know good routes we could follow. It turned out that no one had ever had a reason to venture too much farther up into the mountains past their village. Simply growing enough food to survive, and now working on the road, occupied all their energy and time.

After much hand gesturing and a few Polaroid photos, we gained enough of their confidence to be allowed to rent a house for our "hotel" and begin negotiations for help moving our gear. We would have to find the route ourselves, but would have their help carrying food and equipment up the mountain. During our first night, three of our team fell ill with frequent, almost simultaneous bouts of diarrhea and vomiting. Our sad looking little team provided wonder and amusement for the Tongul children, and perhaps added a measure of pity to their early opinion of us.

*Left, above: Tongul woman carrying food home. (PAM)*

*Left, below: Preparing to go to Nun Kun at our rented home in Tongul. (PAM)*

*Tongul villagers posing for a portrait on the roof of their earthen home. The villagers were always very cautious as we approached, but soon became curious as we passed out small gifts and took Polaroid pictures of them that they could keep. (PAM)*

## SHRINKING WATER SUPPLY

WHILE THE NUN KUN VILLAGERS DID NOT VENTURE into the mountains, they are entirely dependent on the glaciers for their water supply. When the seasonal monsoon ends, it is meltwater from the mountains that keeps the water flowing the remaining eight months of the year. The major river systems of India, Nepal, Tibet, and Southeast Asia—the water supply to billions of people—are all fed by Himalayan glaciers. The glaciers are, in turn, supplied by snowfall during the seasonal monsoon, which is preserved through the year by the low temperatures at these high altitudes.

As part of preparing for our expedition to Ladakh, I reviewed all the published reports on Himalayan glaciers going back to the early 1800s. These reports covered 112 glaciers and left little doubt that an alarming number were shrinking, and shrinking even faster in recent decades. That so many people depend on these disappearing glaciers for their water was then, and is now, of great concern, and provided a major impetus for our research.

Our Asian expeditions have since taken us into many other regions, including several expeditions to Mt. Everest, and it is clear from both detailed studies as well as our return visits that glacier retreat is ubiquitous in the Himalayas, and that the rate of retreat has increased dramatically since our first expeditions in the late 1970s. Were the shrinking glaciers a consequence of less precipitation in the mountains or warming? We were eventually to find out that in some places, precipitation was actually increasing, but warming meant more rain and less snow. As a result, year by year many glaciers were starving for lack of snow in the winter and melting.

*Edge of one of the glaciers draining the Nun Kun ice field, Ladakh.* (PAM)

## LOCAL HELP AND ANNUAL LAYERS

THERE WERE NO DETAILED MAPS FOR THE NUN KUN MASSIF and it was completely covered by clouds in the only satellite image I could find. It was shaping up to be a quite an adventure—weeks of travel by bus, truck, and foot to the study site and, once we arrived, a new landscape full of surprises. With the drilling and sampling gear, as well as tents, food, ropes, and climbing equipment we needed the support of local villagers to reach the plateau where the glacier was flat and the snow was cold.

Life is extremely hard for these people—subsistence agriculture, short life spans, and isolation throughout much of the year. Their lives were completely regulated by the passage of the monsoon. I hired 40 villagers to help with the loads and hoped that they might have some knowledge of the higher terrain to aid us in finding routes, but it turned out they had little need to go to altitude and we spent several days looking for routes. The best route required climbing up three, several 100-foot high icefalls. In addition to having to find routes to the high plateau where we would do our sampling, we also had to find a way to distinguish each year's snowfall from the other. This was critical to using the ice core record to understand the past history of the monsoon.

As we climbed the last icefall, we found the key to dating our core, literally, in front of us. On the face of an ice cliff we saw thin, dark horizontal bands. Spring dust blown off the Tibetan Plateau from the north had left an easily distinguished seasonal mark and we could determine the age of the snow samples by simply counting the dust bands. Even before bringing the samples back to our lab for analysis, we already learned a great deal about the monsoon's history just by observing the thickness of each layer, with a thick annual layer implying that the monsoon moisture had penetrated farther north than usual. It was a great discovery for us—a technique we, and others, have used in dating and interpreting the climate record of ice cores ever since.

To keep the icefall route safe and our supply line open, we had to reset the ropes every morning as the summer sun from the previous day melted out our ice screw anchors. By the afternoon, avalanches were common enough that we stayed off the route altogether.

In planning the expedition, I had purchased enough heavy-duty sneakers for all of the porters and mountain boots for those who would work in the highest areas with us. Though I handed out this footgear, none of them used it. They all chose to go barefoot over rock and ice, preferring to keep their shoes and boots new. New, they would bring a high price in the bazaars—cash that would allow them to buy essential supplies for the long winter. Seeing their world and how different it was from ours was intensely sobering.

Their help carrying our gear and supplies up the icefalls was critical to our success that year. In return, we paid them as agreed, and perhaps just as importantly, gave them every possible item we could leave behind, including our climbing ropes, cans, clothes, and shovels. Among the parting gifts was a Polaroid picture of each porter—a greatly valued item. It turned out, however, that they all wanted to be photographed wearing our exotic western clothes. So one by one, each of the 40 villagers who worked with us took a turn being photographed wearing my clothes. With barely one set of clothes left to each in our team, and my one set significantly used, we said goodbye to the people of Tongul with whom we shared no common tongue, but from whom we had learned about lives deeply dependent on even small changes in climate.

*Several of the 40 villagers who helped us get our equipment onto Nun Kun. (PAM)*

*An exposed cliff edge of the Nun Kun Plateau reveals the dust bands that allowed us to annually date the ice core in the field, followed by validation using chemical measurements back in the laboratory.* (PAM)

*Some of the Tongul villagers we hired to carry loads, waiting as expedition plans are finalized.* (PAM)

*Villagers who helped carry our gear onto Nun Kun, resting in the upper reaches of the icefalls.* (PAM)

*Ascending one of the Nun Kun icefalls. The bumpy snow in the foreground is the result of avalanches, which we avoided by climbing in the morning before the warmer temperatures of midday.* (PAM)

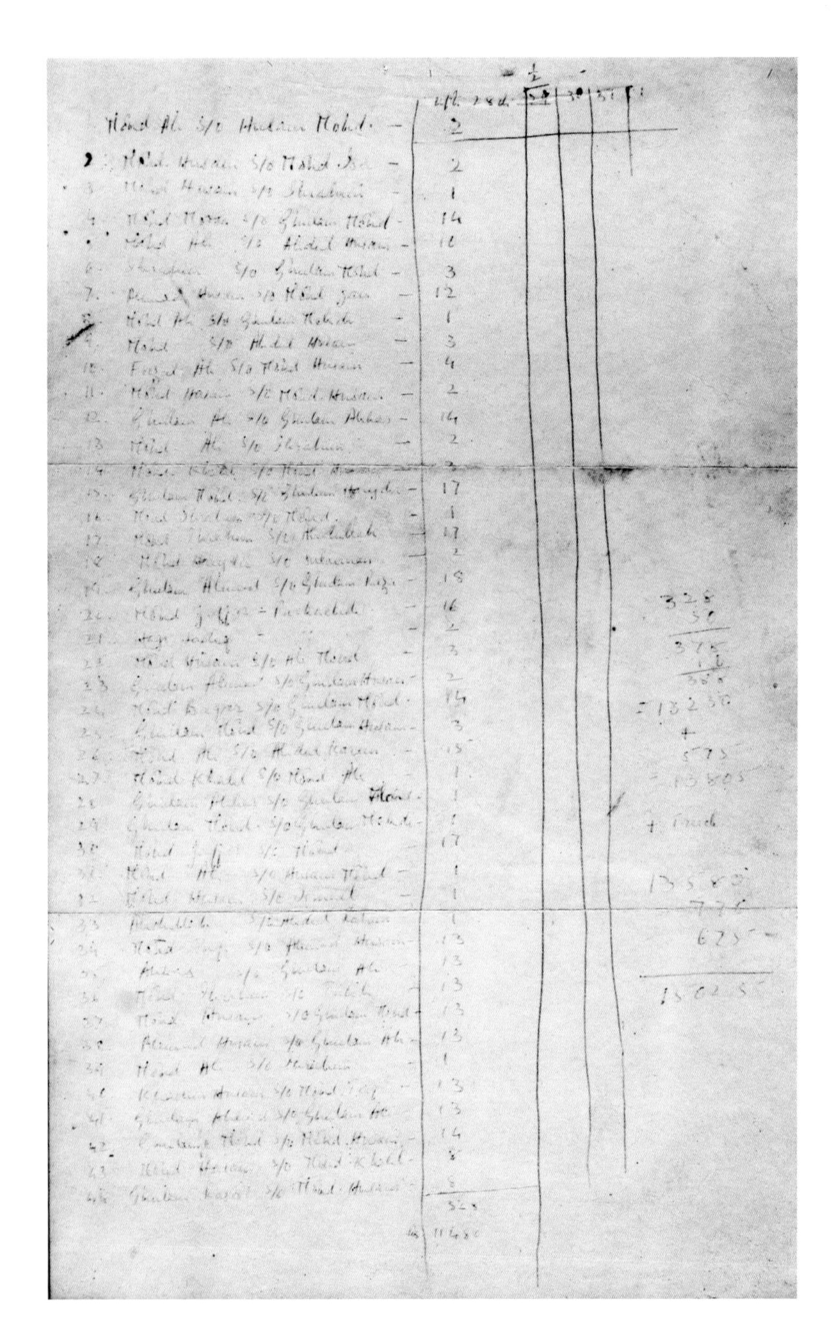

*A list of names of the Tongul villagers who helped us with our expedition.* (PAM)

*Right: Approaching the Nun Kun icefalls that we ascended to reach the plateau.* (PAM)

## FINGERPRINTING THE WIND

THE ROUTE BETWEEN OUR NUN KUN BASE CAMP at 12,000 feet and our sampling camp above 18,000 feet traversed three icefalls. At the high camp we recovered several ice cores. Every time the drill was hauled to the surface we had an additional 40 inches or 50 inches of record. Near the bottom of the drill hole, as drilling became very difficult, it was closer to 5 inches to 10 inches at a time. With each hand rotation of the core, and every time we pulled the drill up out of the hole, we were reminded that we were living in a place that had half of the oxygen in the air that we were used to at home. Each piece was catalogued to identify dust layers and then, while wearing ultra-clean suits over our tattered clothes, we carefully scraped the outside to remove contamination. We placed each cleaned section in pre-cleaned plastic bags and, when we had enough bags to fill a backpack, transported them to the low camp where we melted and bottled them for the journey back to our laboratory for analysis.

Our drill was stainless steel, heavy, and entirely manual. By today's standards it was inefficient and awkward. Though a far cry from the 10,000-foot ice core we would recover a little over a decade later in Greenland—the longest available from the Northern Hemisphere—our 56-foot ice core covering just under 20 years of record was quite a feat for the time. I was excited by our success and our discovery of the annual dust bands. I hoped our analyses would tell the history of the monsoon year-by-year. As it turned out, we were able to track changes not only year by year, but also season by season—an ability that was to revolutionize the field of climate change.

Serendipitously, close to the end of our time on the plateau, I watched as clouds from the south, from the Indian Ocean, rose up from the valleys, and clouds from the north, from over the Tibetan Plateau, also moved in—both part of the seasonally reversing monsoon that dominated the region and the lives of all its inhabitants.

As soon as the snowstorm passed, I immediately started down our now-familiar icefall route, collecting samples from 18,000 feet all the way to 14,000 feet, being sure to sample the two air masses, higher snow from the north, and lower snow from the south, and their intersection. The results of this impromptu study showed clear differences in the chemistry of the two air masses, and provided a "dictionary" that greatly aided in reading our ice core histories.

Our analyses also held a few surprises. We found higher levels of sodium and chloride from sea salt and evidence of warmer temperatures in the snow from the southern clouds, and higher levels of dust and evidence of colder air in the northern. In the southern snow we also found ammonium and phosphate—chemicals that could only come from the use of fertilizers below us and to the south in India. I remember the day at our lab when the data started coming in, realizing that ice cores could yield fingerprints telling where the air had come from, and what processes were occurring at the source. They offered remarkably accurate and detailed insight into human activity and—as hoped—into the historical behavior of the monsoon. That day's insight revolutionized our thinking and laid the foundation for many future studies.

*Evening approaches on the Nun Kun Plateau. (PAM)*

*"Banner" clouds on the lee of Himalayan summits to the south of our Nun Kun camp. These clouds were telltales of warm, moist air moving north from the Indian Ocean.* (PAM)

*Left: Our camp halfway up the route to the Nun Kun drilling site. We would often stop here, waiting for the cooler parts of the day to ascend the icefalls without the danger of avalanches.*

*Right: Ice coring on the Nun Kun Plateau was a slow process. With less than half of the oxygen at sea level at this coring site, raising the heavy stainless drill system was a challenge. Our camp is seen below.* (PAM)

RESEARCH GROUP
DEPT OF EARTH SCIENCES

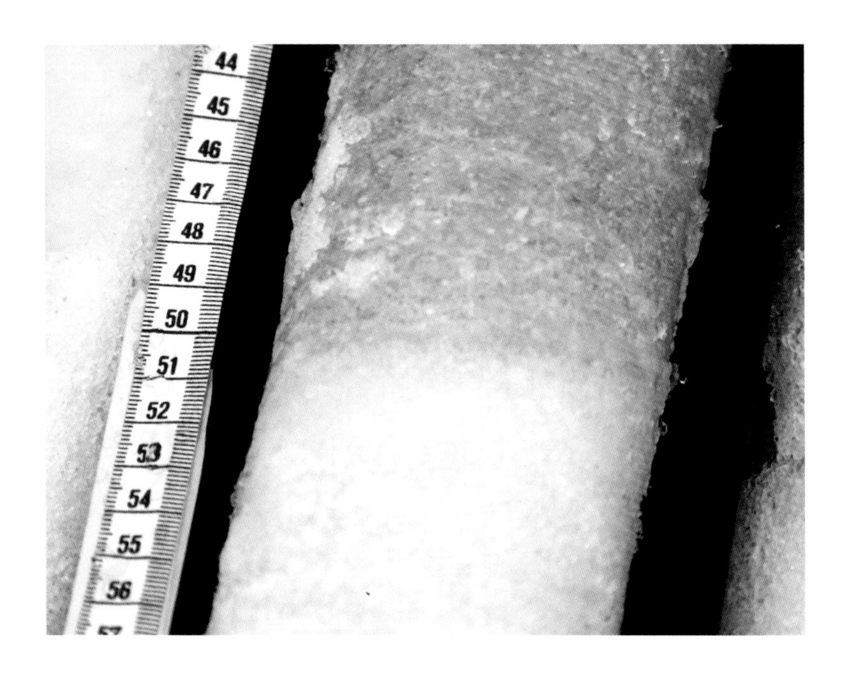


*An ice core from Nun Kun showing the spring season dust bands we used to determine annual layers needed to date the core.* (PAM)

*Left: Berry Lyons (now Director of the School of Earth Sciences at Ohio State University) melting ice samples recovered from the Nun Kun Plateau and bottling them at one of our lower camps.* (PAM)

*Right: Peter Axelson (then a University of New Hampshire undergraduate) sampling snow on a high ridge near Nun Kun.* (PAM)

## IS HUMANITY THE CAUSE?

IN THE SPRING OF 1984, OUR RESULTS FROM NUN KUN were published showing, among other observations, that the snow was more acidic than expected. I was used to the scientific discussions that often followed the publication of a new paper, but this one was different. The day it was published, phone calls began pouring in from a wide range of journalists, environmentalists, industrialists, advocates, and lobbyists.

In 1970, the U.S. Congress had passed the first Clean Air Act (strengthened in 1977 and again in 1990). Though the legislation passed, many still argued that industrial emissions did not—could not—have a perceptible impact on a hemispheric, or even regional scale, just as some argue today that greenhouse gases do not have an effect. In this setting, advocates of both positions were calling me, seeking proof of their position in our results.

By this time, I was sure that humanity could have, and was having, a significant impact on planetwide processes, including acid rain, but our Nun Kun results could not demonstrate whether the acidic snow was the result of industrial emissions. To do this, we would need ice cores that provided records extending hundreds of years into the past—a core we would soon collect from near the Dye 3 station in southern Greenland showing not just the rise of industrial pollution worldwide, but the effect of the Clean Air Act on reducing it.

But on that spring day and during the following weeks, I tried to explain to callers what we could and could not tell from our acidic snow. I don't know how many understood, and I don't think any of them liked what I had to say—that we needed a longer perspective.

*Full moon over Nun Kun. (PAM)*

CHAPTER FOUR

# EARLY YEARS IN GREENLAND

*A C-130 aircraft delivering supplies to the Greenland Ice Sheet. (MCM)*

## GOING TO GREENLAND

WHILE DEBATE CONTINUED OVER THE CLEAN AIR ACT, and whether human industrial activity and pesticide use were having a significant effect, I had begun to think we might be able to provide ice core data to help answer the question. By the early 1980s, the ability to measure the chemistry—even very low levels in ice cores from remote areas—accurately enough to evaluate industrial emissions was a reality, and we had been honing our skills. What I had in mind was to collect a long, comprehensive history of atmospheric chemistry extending to well before the industrial revolution and into modern times. I hoped we might show whether industrial pollution had traveled out of the population centers of Europe and North America where it was created, and whether industrial emissions had significantly changed the chemistry of Earth's atmosphere. We jumped into the debate with our ice coring and expedition gear.

In late June of 1984 we flew into Sondrestrom Air Force Base (now Kangerlussuak), readied our gear for the field, boarded a ski-equipped C-130 aircraft, and flew to a site known as Dye 3, part of the Cold War Distant Early Warning system, in southern Greenland. As we flew in low, up the deep, steep-sided fjords and valleys, I imagined this place just 15,000 years ago when ice filled these valleys to their brims—well over our flight altitude. Though we were in pursuit of industrial pollution, I found myself thinking of the difference a few thousand years and 15 °F to 20 °F had made: the loss of a continent-sized ice sheet, 300 feet of sea level rise, and the ability to fly a four-engine aircraft where miles of ice had so recently stood. I still remembered the observations I had made in Antarctica revealing a recent retreat of several coastal glaciers. I wondered: What if they weren't anomalies? What if some climate processes moved much faster than we thought?

In the end, Greenland cores answered questions about the effects of industrial activity and policy responses to it, of climate volatility, and much, much more.

*The edge of the Greenland Ice Sheet from the air.* (MCM)

## CLIMATE VOLATILITY

IN THE 1970s, I LEARNED THE PREVAILING THINKING of the time: Climate changed slowly, taking thousands or more years to respond to controlling factors. We knew that there had been a global glacial cycle every 100,000 years or so for at least the last million years, with roughly 90,000 of them cold and 10,000 warm. We knew that we were currently in the warm part of the cycle and had been for about 10,000 years.

But a Greenland ice core retrieved in the early 1980s by a joint Danish, Swiss, and U.S. team revealed an intriguing glimpse into the true, volatile nature of Earth's climate. In this core, there appeared to be a large, long-lasting change in temperature, over the unimaginably short period of 70 years about 11,000 years ago. The location of this core—logistically convenient, but not a perfect ice coring site—might mean that this apparently rapid change was simply a result of irregular ice flow and disturbances in the record rather than a true reflection of climate.

In the early 1990s, I would organize and lead one of the two coring efforts at the best sites in Greenland that, together, confirmed that this rapid change in temperature was in fact a true insight into climate change and not an anomaly of the coring location. We were to discover that high- to mid-latitude temperatures could go up or down as much as 25 °F in as little as two years, often in under a decade, and be sustained at the new level for decades to millennia. I was to discover with my colleagues that these changes were produced by abrupt shifts in the position and strength of atmospheric circulation patterns, such as the jet stream that travels westward across North America and Eurasia. Such events must surely have posed enormous challenges for humans and many other species, and I began to wonder if they could happen today.

For my part, when I first learned about the possibility of a major climate change event occurring in as little as 70 years, I thought again of my Antarctic glacier retreat studies suggesting a more rapid, modern process, and of my study of Himalayan glaciers hinting that they were retreating faster in recent decades. Doubts about the gradualist paradigm began to gain a serious hold in my thinking, though I still thought of these events as fleeting—like a few days' rise in a falling stock market.

*Valleys on the coast of Greenland that just 14,000 years ago were filled with a mile or more of ice.* (MCM)

*Evening drilling in southern Greenland. (PAM)*

## HUMANITY IS RESPONSIBLE

IT WAS THE SUMMER OF 1984 and our C-130 aircraft banked, descended out of the clear Greenland sky, and landed at 8,100 feet above sea level on the packed snow of the Dye 3 airstrip amid a swirl of prop-washed snow, 24-hour daylight, and -20 °F temperatures. Dye 3 offered a convenient destination because we could piggyback on the military flights already going there to support the station.

The six of us unloaded our gear and two snowmobiles, drove to a site 20 miles upwind of the station, set up our Scott tents, and began ice drilling operations. The record of industrial emissions we hoped to collect was then at the edge of current technology, and we went to great lengths to ensure that the core would not be contaminated. One of our main concerns was how to power the drill needed for this core. Typically it was powered by a gasoline generator, but such a generator would produce many of the very compounds we hoped to measure and would likely compromise the ice core record. The solution came from an innovation developed by a technical colleague, Bruce Koci. He devised a drill system powered entirely by solar panels, providing the clean conditions our study required. Over a few weeks of steady work in the high, cold world of the Greenland Ice Sheet, we extracted 300 feet of pristine core—a core that provided a record of atmospheric chemistry back to the year 1767.

Our results, published in 1985, showed pollution levels rising in perfect step with human industrial activity: the reduced emission levels during the Great Depression, the increased levels restarting with the postwar boom in the late 1940s and, significantly, the reduced levels following passage of the Clean Air Act even during economic expansion. The data could easily have been mistaken for one of global economic output until the Clean Air Act, where pollution leveled off and no longer tracked economic activity.

We had definitively shown that industrial sulfate and nitrate emissions, the two components of acid rain, had a hemispheric impact and that the Clean Air Act legislation—like the Montreal Protocols that would come in 1987—could lead to significant reductions, even during strong economic expansion. I was elated by the clear result we had been able to achieve: to have been able to end any serious debate on the global importance of human industrial activity, to have demonstrated the significance of policy responses to it, and to have provided scientific support for strengthening the Clean Air Act in 1990.

*Solar-powered ice core drilling in southern Greenland. (PAM)*

## MORE THAN ACID RAIN

THOUGH IMPORTANT, ACID RAIN WAS NOT the only human impact revealed in our cores. It was becoming increasingly obvious from our ice core records that human industrial activity was resulting in dramatic increases in sulfate, nitrate, lead, radioactivity, and much more, and that the age of human-caused, major environmental changes was well under way, along with the disease and ecosystem destruction they caused.

A notable illustration is the use of lead, a potent poison, in automotive gasoline. Building on advances in collecting ultra-pristine ice cores, Clair Patterson of Caltech applied his extraordinary analytical skills to measuring the lead in Antarctic and Greenland ice cores. He showed that, like other pollutants, lead in the global atmosphere had risen with the use of gasoline, and fallen dramatically with the ban of leaded fuels. Beginning in 1965, he tried to draw public attention to the problem of increased lead levels.

There was even more to the acid rain story. By the 1970s, science had a record of global average temperature measurements beginning in the late 1800s. This record showed rising temperatures until the 1940s, after which they stopped rising. At the time, some hypothesized that the slight cooling we saw after 1940 might be the beginning of the end of the current, 10,000-year warm period—and the beginning of a slow, many thousand-year return to a colder time. Though of academic interest, we thought that even if this were true, the gradual cooling could not possibly be of practical importance for humanity in the foreseeable future, because climate operates so slowly.

Unlike carbon dioxide and chlorofluorocarbons (CFCs) that remain in the atmosphere for hundreds or thousands of years, sulfate, being water-soluable, is removed relatively quickly by rain and snow—usually in less than a year or two, depending how high in the atmosphere the sulfate reaches. Volcanoes emit sulfate, often in very large amounts and sometimes high up into the atmosphere, and we found "spikes" of sulfate associated with volcanoes that were often many times higher than the usual amount in the atmosphere. Because the volcanic events were short-lived, global levels of sulfate returned to typical levels within a few years. The high and growing levels of sulfate in the atmosphere seen in our Greenland ice core meant that there was a large, ongoing source: industrial emissions.

In reviewing the volcanic spikes, we noticed that regionwide temperatures were consistently lower when sulfate was high. This made sense since sulfate was expected to reflect sunlight and thus reduce the amount of heat that reached the Earth's surface. But we now had a record showing that this was in fact the case, and published a paper in which we hypothesized that the cooling observed between the 1940s and 1970s was not—as we had naively guessed—a harbinger of a new ice age, but the cooling effect of industrial sulfate.

This hypothesis has since been validated as one of the primary controls for cooling in the Northern Hemisphere between the 1940s to 1970s, and represents a critical aspect in our understanding of Earth's heat balance. By the late 1950s and early 1960s, we thought that emissions of carbon dioxide should have been increasing the amount of heat kept at the Earth's surface. We did not then have the extensive oceanic monitoring programs in place to tell how much heat was going into the oceans, which have absorbed the vast majority of global warming heat because water holds so much more heat than air. Further, looking at the 1970s, the atmosphere seemed to be cooling, or at least not warming up. So researchers studying carbon dioxide's effects were unable to connect theory with observations. By the time the Clean Air Act was passed and we began to reduce atmospheric sulfate, carbon dioxide had built up to much higher levels and dramatic warming became evident in the oceans and the atmosphere.

When we published our hypothesis in the mid-1980s, I had come fully into the understanding that human activity was not only significant on a global scale, but had become a major competing force with natural controls on the climate system.

*Bruce Koci (then at the Polar Ice Coring Office, University of Nebraska, Lincoln) standing in the trench with the solar-powered ice coring system. (PAM)*

CHAPTER FIVE

# THE SEARCH INTENSIFIES

*Our camp at the head of the Beardmore Glacier in Antarctica. (PAM)*

GROWING UP IN THE 1960S AND 1970S, I (Michael Morrison) was fascinated by the full range of Earth's processes: the formation of mountains, rivers, and oceans; the endless variety and beauty of living creatures and ecosystems; the astronomical and cosmological setting in which we find ourselves; and the fascinating twists and turns of human nature and Civilization that has sprung from it.

In the early 1980s, I took a wide range of graduate courses seeking a department that would satisfy my curiosity, but while I found each intriguing, I inevitably felt a frustration with the narrowness of any given course or department. It was then, in spring 1985, while eating lunch in a cafe in Durham, New Hampshire, that I picked up a local paper and read about a new research initiative and graduate program at the University of New Hampshire: Earth System Science. This program, the article said, was designed to meet the emerging challenges of answering questions that crossed many disciplines. In order to understand large-scale Earth processes—such as the comings and goings of glaciers; the evolution of large ecosystems, such as the Arctic tundra or Southern Ocean phytoplankton; or the impacts of human activity—we needed to bring together insights from astronomy, biology, oceanography, geology, physics, anthropology, and more. This program would be housed in a new building and would include scientists from many disciplines. I went that day to see how I might get involved.

I became the first graduate student in the new program, studying the emissions of biologically generated sulfur gases from salt marshes with Professor Mark Hines. These gases, the ones that make salt marshes smell sulfury, result from a complex interaction among sediments caught by marsh grasses trapping dead vegetation, microbes that decompose them in the absence of oxygen, and the dissolved sulfate normally present in the brackish water. The gases were of interest because they can play an important role in "seeding" clouds—that is, turning invisible water vapor into visible clouds. On a global scale we wondered if they might contribute to the overall amount of clouds in the atmosphere and, thus, how much sunlight reaches the Earth's surface. With its broad scope and potentially global significance, it was an exciting project.

By the summer of 1988, I had finished my master's in Earth System Science. One of my committee members, Professor Paul Mayewski, directed the Glacier Research Group and my mountaineering interest had led me to learn more about his expeditions to the Himalayas, the Arctic, and the Antarctic. I knew that he was planning an expedition to Antarctica, departing in October. At the time I was considering what I might do next and wondered how I might get involved in some of these cold-region projects.

Luck was with me that fall. Julie Palais, a research professor at the University of New Hampshire, was also going to Antarctica in October and needed a field assistant. "Would I be able to go?" she asked. I jumped at the chance and spent the next few months preparing to help Julie in Antarctica's extreme conditions. Julie and Paul's expeditions would dovetail at a sampling site on the Newall Glacier, so I might also help Paul with his efforts.

A few days before our departure, I noticed a piece of paper taped to Paul's office door: written diagonally across the page, in large, Magic Marker letters was the word, "FUNDED". The page was a letter from the National Science Foundation (NSF) saying that the second Greenland Ice Sheet Project (GISP2) had been funded. The taped sign made it clear that this was a significant moment, but the full significance would only be revealed in the months and years to come.

*Processing ice cores on the Beardmore Glacier, Antarctica. Clean suits ensure that we do not contaminate the core. Photo by a team member.*

*Left, One of the ultra-clean Teflon chambers we used to measure sulfur emissions from New Hampshire salt marshes. The white object to the right of the chamber is a liquid nitrogen bath that traps collected gases.* (MCM)

## NOT AS ISOLATED AS WE THOUGHT

IT WAS DECEMBER OF 1988, and we sat in relative comfort for the three-hour flight from McMurdo Station to South Pole Station in a ski-equipped C-130 Hercules. As we flew over the Beardmore Glacier, I thought about Scott's 1911 expedition, just as I had in earlier days when I traversed the Transantarctic Mountains. It took Scott nearly three months of extraordinary effort under extreme conditions and with great risk. Our flight path roughly followed their route, beginning at McMurdo Station near Scott's Hut, and crossing over the Transantarctic Mountains at the Beardmore Glacier. At 77.5° south, McMurdo station is hardly tropical, but stepping out of the plane into the frigid, high-altitude air of South Pole after leaving sea level McMurdo just three hours earlier was staggering.

We had come to the pole to dig a snowpit about 20 feet deep with steep walls that we would sample very carefully every half inch and produce a proxy record of high-altitude ozone extending back before monitoring had begun. Our results confirmed that the dramatic ozone loss scientists had been observing at the time was in fact a unique, modern event. But there were other surprises in store for us. Having collected the snow samples, we performed many analyses on them and found radioactive material from the Chernobyl nuclear accident two years earlier. From this we learned that not only did pollutants from areas in the Northern Hemisphere like Siberia, where the Chernobyl nuclear power plant had exploded, make their way to Antarctica, but because we were able to date the snowpit samples so well, we also learned just how long it takes to make the trip through the high atmosphere from Siberia to Antarctica: about 18 months. We also found evidence of tropical Pacific Ocean air masses that had penetrated as far as the South Pole: a source we did not think, until then, had any impact on Antarctica. The thinking I had learned in college, that Antarctica was unchanging and isolated from the rest of the planet, finally crumbled completely.

Four of us—my graduate student Cameron Wake from the University of New Hampshire, Barry Lopez (of *Arctic Dreams* fame on a National Science Foundation writer's grant), and my new friend and colleague Michael Morrison—spent a week digging and sampling the 20-foot snowpit 26 miles upwind of the South Pole Station. The temperature at the bottom of the pit hovered around -50 °F, and the hours we spent at the bottom, carefully sampling with our thin, ultra-clean gloves, with little exertion to warm us, made for cold work indeed.

Though the snowpit work was consuming, I was also thinking about the large drilling project, GISP2, that had just been funded a few months earlier. I was to lead this project that would ultimately recover a detailed, 110,000-year history of Northern Hemisphere climate from a 10,000-foot ice core drilled at the highest point of the Greenland Ice Sheet. GISP2 was designed to make a wide range of analyses from each layer of ice and the participation of over 25 research institutions was anticipated. The coordination effort would be significant—more so because European researchers were planning a separate, companion coring effort that would complement ours, and because our first field season was slated to begin in April of 1989—just a few short months away. As we rested on our breaks, I began talking with Michael about what was needed, how we might meet the challenge, and whether he might help me make the project a success.

Having completed our sampling, we radioed South Pole Station and were picked up by a small, tracked vehicle called a Sprite. On one of our earlier radio check-ins, I had made a joke about ordering take-out pizza, being about as far from a pizza place as one could be anywhere on Earth. To our immense surprise and delight, when the Sprite arrived to pick us up, we were greeted with a hot pizza—prepared and sent out to us in an insulated container by the South Pole Station cook. My mind reeled with the collision of worlds at that moment. We had been living in Scott tents, experiencing the same temperatures 24 hours a day as Scott's team had, and had determined our position with a sextant, much as Scott had done. And yet at the same time, we were enjoying hot pizza ordered via radio on our way back to the safe South Pole Station in a warm, motorized vehicle. While I was enjoying the pizza, I was also thinking how glad I was that Antarctica was still vast, still merciless, and still offered plenty to explore.

*View of South Pole station from the cockpit of a C-130. (MCM)*

*Attaching our loaded sleds to the Sprite that would take us from South Pole Station to our snowpit site.* (MCM)

*Paul Mayewski riding one of the sleds as we left South Pole Station. We stopped every quarter mile to place a bamboo flag marking our route.* (MCM)

*We kept collected ice cores as cold as possible before transporting them back to our laboratory. Here, we are digging a storage trench on the Newall Glacier, adjacent to the ice-free valleys of the Transantactic Mountains. Though the air temperature was cold, the sun was bright at this high altitude. Sheltered from wind in the trench and working hard to remove the snow, we became very warm indeed. Left to right: Paul Mayewski, Barry Lopez, and Berry Lyons. Photo by a team member.*

*Left: Team members sampling near the bottom of a 20-foot-deep snowpit 25 miles from South Pole. Clean suits prevent contamination. This snowpit revealed evidence of the Chernobyl nuclear accident and the unique state of the modern Antarctic ozone hole, and demonstrated that tropical air masses can reach far into the Antarctic Plateau. Photo by a team member.*

*Our camp on the polar plateau. (PAM)*

*Right: By digging two snowpits next to each other and leaving just a thin wall between them, we could cover the top of one and observe the snow layers with sunlight coming through from the other. This photo shows the beautiful blue light we experience in these studies, as well as the annual layers used to interpret the age of the snow. (PAM)*

BY THE 1980S, ICE CORE RECORDS WERE RECOGNIZED by the scientific community as a powerful and unique tool for understanding past climate. With the Nun Kun ice core, we discovered a variety of human and nonhuman, and regional and global climate histories captured in the core: the back and forth of the salty monsoon and dusty Tibetan Plateau air masses, the use of fertilizer, and hints of the global reach of acidic industrial emissions.

By the late 1980s, we had produced records of industrial pollution; human economic activity; legislative successes; nuclear bomb testing and accidents such as Chernobyl; solar activity; temperature; the amount of polar sea ice; atmospheric and oceanic circulation patterns; greening and desertification of land areas; forest fires; and volcanic events. Among histories produced by others was the remarkable, 450,000-year record of greenhouse gases—carbon dioxide and methane—from the Russian-French-U.S. ice core at Vostok Station in the interior of East Antarctica—a record that demonstrated the long-term, historical link between these gases and the amount of heat trapped at the Earth's surface. We began to recognize that ice sheets and glaciers were the world's greatest climate library, waiting for us to read the histories.

In this setting, the wide range of factors that contribute to climate change, and the local and planetwide events recorded in any given ice core were becoming apparent. One small moment that drove this point home to me occurred when we were preparing samples from a core we collected on the Beardmore Glacier, the large outlet glacier that Scott and his team traversed on their epic South Pole journey. As we melted samples for analysis, pieces of Styrofoam bounced to the surface from the more recent ice. The only possible source was Styrofoam then in use at the South Pole Station, nearly 300 miles upwind from our study site. This Styrofoam was now a part of the regional climate around South Pole, and our core revealed the time its use began. A few years later, the U.S. stopped any shipments of Styrofoam to Antarctica, and the same Beardmore core, if taken again today, would show a "Styrofoam layer," revealing the period when it was in use at South Pole.

It had become clear to me that in order to reconstruct global climate, we would need a global array of cores. Such an array would provide a picture of global and local climate, the effects of human activity, and clues as to how the many parts of the Earth's climate system work together. I decided to focus my efforts going forward on producing the ice core records needed to make this picture.

Along with many smaller efforts, I would lead two major projects. The first would be the Greenland Ice Sheet Project Two (GISP2), with a team of 25 U.S. institutions, which would collect a detailed, 110,000-year record from a 10,000-foot core drilled at the highest point of the Greenland Ice Sheet in the early 1990s—the longest detailed ice core record available in the Northern Hemisphere. The second would be the International Trans-Antarctic Scientific Expedition, or ITASE, that now includes 21 nations collecting more than 100 ice cores and other snow samples from across Antarctica. On our other expeditions we would recover ice cores from along the spine of mountains reaching from the Andes to Alaska, and from the Himalayas and Tibetan Plateau.

*Using a sextant to determine the latitude and longitude of our sampling site many miles away from the South Pole. (MCM) Inset: Paul Mayewski checking the flag line back to South Pole Station. (MCM)*

## LUSH AND FRAGRANT

IN OCTOBER OF 1988, I (Michael Morrison) stepped off a plane into the New Zealand summer. It seemed familiar, albeit more pastoral and uncrowded than my New England home. I was accompanying Julie Palais (now the Glaciology Program Manager at the Office of Polar Programs, National Science Foundation) as a field assistant, and planned to join Paul Mayewski in the field and assist his team with a snowpit at the South Pole.

From Christchurch, a C-141 took us to McMurdo, a flight without interim landing spots and, once past the halfway mark on fuel, is bound for a landing in Antarctica regardless of developments in the weather. We landed smoothly and soon began our research tasks. Before joining Paul's team on the Newall Glacier, Julie and I collected snow samples to measure trace metals that had made the same trip to Antarctica simply by riding the world's wind currents. We traveled to the shoulder of Mt. Erebus, the world's southern-most active volcano, put on clean suits, and collected samples wearing surgical masks and thin, ultra-clean gloves that protected the snow from contamination, but did not protect our hands from the cold. Having finished our sampling, I found myself staggered by the silence that surrounded us. Without wind, there was no sound but my breath and the blood in my ears. I was fascinated.

We next set up camp on the Newall Glacier above the Dry Valleys of the Transantarctic Mountains, across the bay from McMurdo Station—a few Scott tents a mile downhill from our next sampling site. As with the Mount Erebus site, the silence was profound, but here I also began to notice a complete lack of smells, except for the food and fuel we brought with us: an olfactory silence. There is no liquid water on these glaciers and ice sheets, and thus life is extremely limited. Any slight smell that might come from the exposed rock of nearby mountains was too faint to register. Antarctica was providing me with an unfamiliar and captivating sensory experience, one that took a further turn at our sampling site upwind from the South Pole a few weeks later.

Traveling the 26 upwind miles to the study site with Paul and colleagues, we followed a course determined by distance and azimuth, setting bamboo flag poles every tenth of a mile or so to mark the way. Bouncing over small snow dunes known as sastrugi, riding on the sled carrying the poles felt more like being on a boat on the ocean than the frozen desert of the Antarctic Plateau. Maintaining complete skin cover against the near-instant frostbite of strong winds and -30 °F to -50 °F temperatures, we made our way into the "wilderness."

Though I have always known we live on a rotating planet that orbits a star, standing on the Earth's axis and watching the Sun travel in a perfect circle around the entire sky, I had a felt experience of being on a planet in a solar system for the first time. My place, Earth's place, and our common movement became palpable realities.

I found myself in a world defined by geometry and physics, where the familiar thermal inertia of land and water had no play, and where my location was defined by the ecliptic and the clock, and by angles and distances.

In this setting, I found myself looking for a way to locate myself in time and space. My mind began to travel in ever-widening circles, hunting for historical features like those familiar to me in my North American home: forests that had arrived after the last glaciation; roads, and then highways built over the last century and a half; landscapes shaped by rivers and erosion over the past hundreds and thousands of years; or the cities and towns that had grown in the past several hundred years. Standing on the deep ice of Antarctica was very different. There were no discernible events I could bring to mind, beyond my own presence, at any imaginable point in time. I looked farther and farther back in ever-widening circles. Only when my thoughts reached 35 million years ago did I find a mark—when the ice sheet itself formed as atmospheric carbon dioxide levels dropped below 450 parts per million from the 1,000-plus parts per million of the warm period 55 million years ago. This place, Antarctica, was truly different and expansive in so many ways.

In this silent, immense, and encompassing state, I returned to Christchurch inside a dark-bellied C-130 just six short weeks after I had left. As the door cracked open and then flung wide, a tsunami of odor, heat, and humidity overwhelmed me, like an unwary child in the surf. Standing on the tarmac in the soft evening light, I felt immersed in life. Vegetation, insects, and animals surrounded me everywhere, covering the ground as far as I could see, making homes even in the pores of brick and stone, and the bottoms of streams and puddles, filling the air with sounds and smells—relentlessly, inescapably, filling my nose, eyes, and ears.

Vigorous, ubiquitous, and inexorable, life seemed everywhere, fully occupying the landscape. The signs of human history and events that had occurred in the fantastically, unimaginably, short time spans of hundreds of years—and even months, days, and hours—were everywhere.

The world was suddenly close again. I had anticipated having to adjust

upon arrival in Antarctica, but had not thought I would have to adjust upon returning to Civilization. As I underwent this unanticipated acclimatization, I realized that in Antarctica, I had become attuned to an "extraterrestrial" mind—a mind of eons, stars, and physics, and that my return to New Zealand was like visiting our strange, bustling, and burgeoning home for the first time. Not only had I had the privilege of visiting Antarctica, but also of "returning" to Earth and deeply feeling its living wonder.

*New Zealand farms near the Abel Tasman National Park.* (MCM)

CHAPTER SIX

# FAST CLIMATE

*The 52-foot-diameter geodesic dome that held the 100-foot-high GISP2 drill, and the surrounding field camp.* (MCM)

## CROSSING GREENLAND

BEGINNING IN 1985, I led a series of over-snow traverses on the Greenland Ice Sheet to collect snow and ice samples that would help determine the best possible site in the Northern Hemisphere to drill a deep ice core. The success of the greenhouse gas record from the Vostok ice core in East Antarctica spurred a keen interest in a long record from the Northern Hemisphere. As part of this effort, we spent a few months each spring for several years traversing the Greenland Ice Sheet—the only ice sheet and the thickest ice in the Northern Hemisphere—to help find the best ice coring sites.

My traverse experiences in Antarctica were helpful as we crisscrossed portions of southern, central, and northern Greenland by snowmobile and Tucker SnoCat. We dug and sampled snowpits, collected shallow ice cores, installed automatic weather stations, and provided on-the-ground fix points for aerial surveys.

Our travels most often started in late spring, so we had the opportunity to feel the changes that come with lengthening days and warming. It was as if the Arctic were coming alive even on this snowy, white, featureless surface. While there is plenty of ice, there is no liquid water —at least at the higher elevations where our studies took us at that time—no soil and consequently no animals or plants. On rare occasions, we might spot a wayward seagull or fox, and there had even been rare reports of coastal polar bears crossing the ice sheet. On the ice sheet, a bear could be expected to be hungry, and we could be expected to be a poor match. Fortunately, bears only appeared in our tent-bound conversations.

*Our team traversing the Greenland Ice Sheet to identify the best site from which to drill the deepest ice core ever recovered in the Northern Hemisphere. (PAM)*

## BUILDING A "TIME MACHINE"

BY 1987, WE HAD IDENTIFIED the best location for a deep ice core in Greenland—the highest point, at about 10,500 feet, and 72° north. With interest still strong, the U.S. National Academy of Sciences and the Office of Polar Programs at the National Science Foundation organized a meeting in Seattle, Washington, to determine how best to proceed with the actual ice core recovery and science program. More than 90 scientists attended and I was elected to lead the U.S. effort—to be known as the Greenland Ice Sheet Project Two (GISP2)—which ultimately included 25 U.S. research institutions.

The National Science Foundation officially funded GISP2 in September of 1988, just weeks before I departed for an Antarctic expedition. The first GISP2 field season was to begin in the spring of 1989 and there was no time to lose. On our breaks from digging the snowpit near South Pole, Michael Morrison and I discussed planning, and I asked him if he would lead the coordination office. He agreed and on January 1, 1989, we moved a desk and chair into a vacant office at the University of New Hampshire where I was a professor and director of a research center, set up a phone and fax line, and GISP2 planning was under way in earnest. In April the first ski-equipped C-130 from the New York Air National Guard made an open-snow landing near the drill site and, with logistics support from the Polar Ice Coring Office at the University of Alaska, Fairbanks, the work of establishing the camp and drilling facility began.

For the next five years, we drilled ever-deeper into the past: back through the Holocene, the most recent 10,000-year warm period; and from there into the 90,000-year—and most recent—glacial cold period. The core was large, 5.2 inches in diameter, and provided ice samples for the dozens of analyses being done by the 25 organizations involved in GISP2.

Many friendships were made during GISP2—friendships that have led to further expeditions and collaborations. The friendship between Michael and me proved invaluable in meeting the many challenges and inevitable unforeseen developments of such a large project. Years later, it would also lead to him joining us on an expedition to the Chilean Andes and to writing this book together.

A companion European ice coring effort, known as the Greenland Ice Core Project (GRIP) was to be located at a site 18 miles east of the U.S. effort and was led by Bernhard Stauffer from Switzerland. Having two cores from different locations near this high point, analyzed by different teams using different methods, allowed us to determine whether the events observed in either core were simply artifacts of the particular coring site and analytical approach, or whether they were indeed true climate events. The results from the two cores were in near-perfect agreement, and provided invaluable assurance that the remarkable climate history our teams produced were in fact well-preserved records of past changes.

The level of detail and the scope of insight into climate revealed by GISP2 and GRIP were unparalleled, and to this day remain a cornerstone in our understanding of climate. Some of the results were available the same day the core was drilled, and some could only be obtained after long, careful lab analyses at our home universities, but the exquisite picture of climate from successively older and older times emerging daily before our eyes left us feeling like we were at the controls of a time machine telling of eras long past. Our time machine stopped when we reached bedrock with a total core length of 10,018 feet and a well-dated history spanning 110,000 years. There were probably another 200,000 or more years of ice in the last few feet above bedrock, but because it was so tightly compacted, it was not possible to recover an understandable history from that period. In 2010 we were to develop analytical techniques allowing us to examine in great detail this highly compressed section of ice. While these results are still forthcoming, we may yet get to extend the detailed portion of the GISP2 climate record beyond 110,000 years ago.

*The first tents at the GISP2 drilling site. Scott tents, like these, and other tents were used as living quarters for most of the field team. (MCM)*

STANDING FOR HOURS ON END IN A SNOW CAVE at -20 °F counting annual layers in an ice core may not sound like a normal summer activity. Neither would attending to every detail of an endless routine of up-and-down drilling cycles, but this, and more, is how the GISP2 team spent every summer for five years. The often tedious work and the efforts of the many people who made GISP2 a reality produced scientific rewards far beyond our hopes and expectations, and ushered in a sea change in our understanding of climate.

Before the Greenland ice cores, we knew that many factors were probably important in determining climate at any point in time. Many of the longer records, like sea sediment cores, could say only what had happened at intervals of hundreds of years or longer and they told about only a few climate controls and responses, so it was hard to tease out all of the important factors and interactions from these coarser records.

By comparison, the Greenland ice cores showed how climate evolved year to year and sometimes season to season. And because of the large array of analyses made on ice from each layer, we could tell how each of the many factors changed together. We identified the interaction of several major influences on Earth's climate: changes in the Earth's orbit around the Sun that result in periodic changes in the amount of sunlight received by the Earth at different latitudes; the amount of ice present on land or melted into the oceans; biological productivity on land and in the oceans; changes in ocean and atmosphere circulation patterns; changes in the output of energy from the Sun; changes in the chemicals in the atmosphere;

*Kris Kingma (then a University of New Hampshire graduate student) cutting sections of the ice core in the GISP2 science trench. (PAM)*

and events like volcanoes, extraterrestrial impact events, dust storms, and forest fires.

The ice core and ocean sediment records demonstrated that, before the industrial era, the big cycles of cold and warm operating over periods of many thousands of years were primarily controlled by small changes in the Earth's orbit, with commensurately small changes in the amount of sunlight. However, these small sunlight changes had a large effect on the level of carbon dioxide held in the atmosphere, oceans, and biosphere, which, in turn, kept more or less of the Sun's heat at the surface of the Earth. Over the past several hundreds of thousands of years, carbon dioxide levels have fallen as low as 180 parts per million by volume. These were the times of the ice ages, times when the Earth was much cooler and northern North America was under as much as two miles of ice. During the warm periods, carbon dioxide reached as high as 280 parts per million by volume, and the great northern ice sheets across North America and Eurasia retreated and disappeared. Changes in the amount of ice and carbon dioxide, triggered by Earth's orbital changes, had been the main factors driving big changes in climate in the past.

The year-by-year resolution of the cores also revealed another important characteristic of Earth's climate—that it has "thresholds". That is, it tends to operate in one fashion even when some important factors are changing, and at some point there is a "straw that breaks the camel's back", when climate changes abruptly, sometimes in just a few years. We were able to verify the climate change thought to occur over a 70-year period—a

*Mark Wumkes, chief driller for GISP2, in the control booth. Mark played a critical role in recovering the deepest ice core available in the Northern Hemisphere—10,018 feet. He also helped us with drilling expeditions in Antarctica and Tierra del Fuego.* (PAM)

*The massive GISP2 ice coring system inside the drill dome.* (PAM)

remarkable discovery from the earlier Greenland core—and, with our much greater detail, found that it actually took place in less than 10 years and perhaps in as little as two years. Changes like this occurred not just once, but several times over the span of our 110,000-year Greenland ice core record, suggesting that, at the very least, the climate of the North Atlantic region could change dramatically and very quickly. With time, abrupt climate change events were identified from ice cores all over the world and even in deep-sea sediment cores as researchers began to probe them more finely. While not all of the abrupt climate change events were necessarily the same in character, we found some common denominators.

Realizing that major changes, often lasting hundreds to thousands of years, started and stopped in as little as a decade, and in some cases as little as two years, made us wonder: What is the key to those changes? The Greenland ice cores answered this question. At each pole there is a somewhat isolated air mass that rotates to the west. These frigid air masses, known as Polar Atmospheric Cells (PACs), routinely expand toward the equator and contract toward the pole with the seasons and everyday weather changes. The dramatic shifts shown in the Greenland cores turned out to be very large expansions and contractions of the northern PAC, accompanied by changes in the extent of sea ice. It was this expansion or contraction that was occurring so swiftly and leading to such dramatic changes in climate. Further, we found that these changes in the extent of the PAC were sometimes associated with dramatic changes in the strength and position of the Gulf Stream, which today brings significant amounts of heat northward and, combined with westerly winds, keeps Europe's climate warm instead of cold like Siberia's.

We found that these changes were largest in the polar regions and smallest near the equator. At the GISP2 site, temperatures changed during the most dramatic of the abrupt climate change events by as much as 30 °F. At midlatitudes, typical of many population centers, temperature shifts were typically in the range of 5 °F to 20 °F. During cold periods, twofold to threefold decreases in average precipitation and strengthening of winds and storms in the mid- to high-latitudes occurred relative to warm periods. The most dramatic of the abrupt climate change events occurred when the ice sheets and glaciers on Earth were at their greatest extent. Over the past 11,000-year warm period, when the amount of ice has been near current levels, climate changes have been relatively less radical than during the previous cold glacial period. It seems that it is no coincidence that Civilization arose during this period of comparative quiescence.

Quiescence, however, is a relative term, well worth considering and respecting. The extremely accurate dates we obtained for events in the Greenland cores allowed detailed comparisons with archeological studies revealing that the much smaller shifts that did occur during the last 11,000-year warm period—very few more than 4 °F—and relatively moderate shifts in storm frequency and intensity proved deadly for many civilizations. The collapse of the Akkadian Empire 4,200 years ago in modern-day Syria and accompanying consequences throughout much of the Middle East into central Asia, and the end of the Maya Empire 1,100 years ago in central America were associated with significant shifts in atmospheric circulation and sea ice extent over the North Atlantic and were recorded in the GISP2 ice core. Those climate conditions led to drought in the Middle East and Central America. The disappearance of the Norse colonies in Greenland 600 years ago accompanied an increase in the amount of sea ice surrounding Greenland and also were found in the GISP2 ice core record. These and many other major disruptions to Civilization and ecosystems are recorded in the GISP2 ice core. If these "small" climate changes had such dramatic impacts, it is easy to imagine that Civilization itself could not have emerged in the far more tumultuous time prior to the last 11,000 years. More importantly for the future, these "disruptions" demonstrate the fragility of civilizations and ecosystems to climate change at any scale.

Because it remains in the atmosphere for hundreds to thousands of years, carbon dioxide is among the most important of the greenhouse gases. It sets the stage for the rapid changes we have observed, and has been responsible for the shifts between cold glacial periods and warm interglacial periods. Except in the case of extensive volcanic activity, such as the one proposed to have been responsible for the Paleocene–Eocene Thermal Maximum around 55 million years ago, carbon dioxide levels in the atmosphere have historically changed only gradually. During cold periods of the past million years, we now know that carbon dioxide levels were about 180 parts per million by volume; during warm periods, about 280 parts per million by volume. In the geologically instantaneous period since the early 1800s, human industrial activity stepped into the historical process by extracting long-buried fossil fuels and adding carbon dioxide directly to the atmosphere, overwhelming any effect that orbital changes in sunlight might otherwise have had. Based solely on orbital cycles, we would predict a mild and slow return to glacial conditions in the coming thousands of years.

In the past 150 years, humanity has put enough carbon dioxide into the atmosphere to raise levels from 280 parts per million by volume to nearly 500. Because the oceans have absorbed about half of our emissions, actual concentrations were near 390 parts per million in 2010. Every year we increase that amount by about 2 parts per million.

We are, as a result, in uncharted territory. Carbon dioxide levels have not been this high for probably 35 million years, when the Antarctic Ice Sheet was just beginning to form. Looking forward, the results from the Greenland ice cores and the work of thousands of scientists around the world have yielded simulations of how things might evolve over the coming decades. What we know for certain is that we are applying a great deal of "pressure" to Earth's climate and, unless we act in a worldwide, coordinated fashion as we did with CFCs, we are certain to pass one—or more likely, several—thresholds and create a climate radically different from the one we now enjoy and which allowed the rise of Civilization. One in which it may be difficult—or impossible—for Civilization to persist in its current state. The ability of Earth's climate to change abruptly, together with the dramatic influence of human activity, means that human-caused climate change will also most likely be abrupt. We would never have expected this without the detail afforded us by ice cores.

The demonstration that abrupt and dramatic shifts in atmospheric circulation, temperature, precipitation, and sea ice extent can occur in less than a few years, and that these changes can be sustained for decades and longer, has immense significance. This is clear from the impacts of these events on past civilizations and ecosystems, but the realization has tremendous relevance to us today too. Prior to the realization that climate could change abruptly, it was assumed that no change save an immense volcanic event or an extraterrestrial impact could disrupt our way of life. Rare, massive eruptions, such as Toba in Sumatra approximately 74,000 years ago, spewed tremendous amounts of sulfate into the atmosphere for a decade or longer and cooled climate for several decades. Typical volcanic eruptions only alter climate for up to two to three years. Large meteor strikes—extraterrestrial events such as the one 65 million years ago—are clearly "showstoppers" yielding major extinctions and landscape changes, but these are extremely rare and certainly do not operate at the repetitive scale of abrupt climate change events.

During some periods in the past, abrupt climate changes occurred fairly regularly—approximately every 1500 years. Over the last 11,000 years, abrupt climate change events have been small and less regularly spaced, but most importantly, they have nevertheless occurred several times.

If, as science assumed prior to the discovery of abrupt climate change, that climate can only respond slowly, then nothing short of volcanic activity or an extraterrestrial impact event would yield a change in climate in less than hundreds to thousands of years. Instead, we found that Earth's climate changes can outpace political terms. Suddenly a new world of change appeared. This meant that what we emitted into the atmosphere could impact climate much faster than previously thought. The climate system was no longer a giant sponge capable of absorbing everything. The climate system could take only so much before it responded, and the response was fast.

The realization that climate can change abruptly immediately raises the question of when the next abrupt climate change event will occur. We understand that climate change is a consequence of several interacting controls—some slow, such as changes in the energy associated with Earth's position relative to the Sun, and some that operate at faster scales, such as changes in energy output of the Sun, natural oscillations in ocean circulation, changes in atmospheric dust and sulfate, and, although usually changing no faster than on the order of hundreds to thousands of years, changes in greenhouse gases.

Today greenhouse gases like carbon dioxide have risen higher and faster than anything recorded in the last million years, and most likely the last 55 million years. The threshold needed to initiate abrupt climate change is now held in human hands through emissions of fast-rising greenhouse gases and sunlight shielding sulfate and black carbon (a product of the combustion of carbon fuels such as coal, wood, and diesel). Sulfate and black carbon are short-lived, unless the sulfate and black carbon emissions are sustained. Greenhouse gases reside in the atmosphere for centuries and so does their capacity to trigger abrupt climate change. We truly have a new state of the atmosphere now—one that has changed rapidly in recent decades as greenhouse gases rise—and it brings with it the near-certainty of a human-caused abrupt climate change.

More about GISP2 can be found in "*The Ice Chronicles*" (2002) by Paul Andrew Mayewski and Frank White.

*The GISP2 ice core was carefully handled to prevent any contamination. Here, electrical conductivity, an in-field measurement that allowed quick identification of acid spikes related to volcanic events, was performed by Ken Taylor of the Desert Research Institute in Nevada. Several other basic measurements were conducted in field laboratories before cutting the core into smaller pieces for transport back to labs in the U.S.* (PAM)

*Looking down from the control booth at the last section of core to come to the surface before the drill reached bedrock on July 1, 1993. Photo by a GISP2 team member.*

*Right: A freshly drilled section of core, 5.2 inches in diameter and 18 feet long, is laid out in the drill dome for initial processing. In order to prevent the borehole from closing in on itself, it must be filled with a fluid that is about the same density as ice. At GISP2, we used butyl acetate. Though a mild chemical, the drill team wore protective suits and the dome was extensively ventilated to maintain good working conditions. After being cut to 3-foot lengths, the core would be lowered to the handling area with a dumbwaiter (in the shaft area at the back center of the photo). (MCM)*

## GREENLAND IS CHANGING FAST

MUCH HAS CHANGED IN GREENLAND since the end of the GISP2 field program in 1993. Satellite observations in concert with ice core records show that large areas of the Greenland Ice Sheet surface are melting for the first time in thousands to tens of thousands of years. Detailed studies along the west and east coasts of Greenland, done by Gordon Hamilton and his graduate students at our Climate Change Institute at the University of Maine, demonstrate massive losses of ice from the east coast of Greenland for at least two glaciers. The melt of those two glaciers alone represents almost 10 percent of global sea level rise for the period 2001-06.

These ice losses are attributed to the invasion of warm ocean water beneath the coastal edges of these glaciers and consequent grounding line retreat. At the same time, warmer air over the glaciers causes surface melting, changing the glacier surface to darker, more heat-absorbent colors, which, in turn, leads to more melting. Finally, in some places the surface meltwater penetrates deep into fractures where it refreezes and expands, resulting in more fracturing and ice breakup.

Of significant concern is the possible effect of fresh meltwater entering the North Atlantic. The Gulf Stream, which brings heat northward and keeps Europe as it is, rather than like Siberia, circulates partly because it is so salty. It is possible that the freshwater from melting Greenland ice could reduce the saltiness enough to slow down the Gulf Stream.

*Gordon Hamilton (Climate Change Institute, University of Maine) monitoring ice loss along the east coast of Greenland. Photo by Leigh Stearns, former University of Maine graduate student, now at the University of Kansas.*

*Evening approaches, marking the close of the Greenland ice core drilling field season.* (MCM)

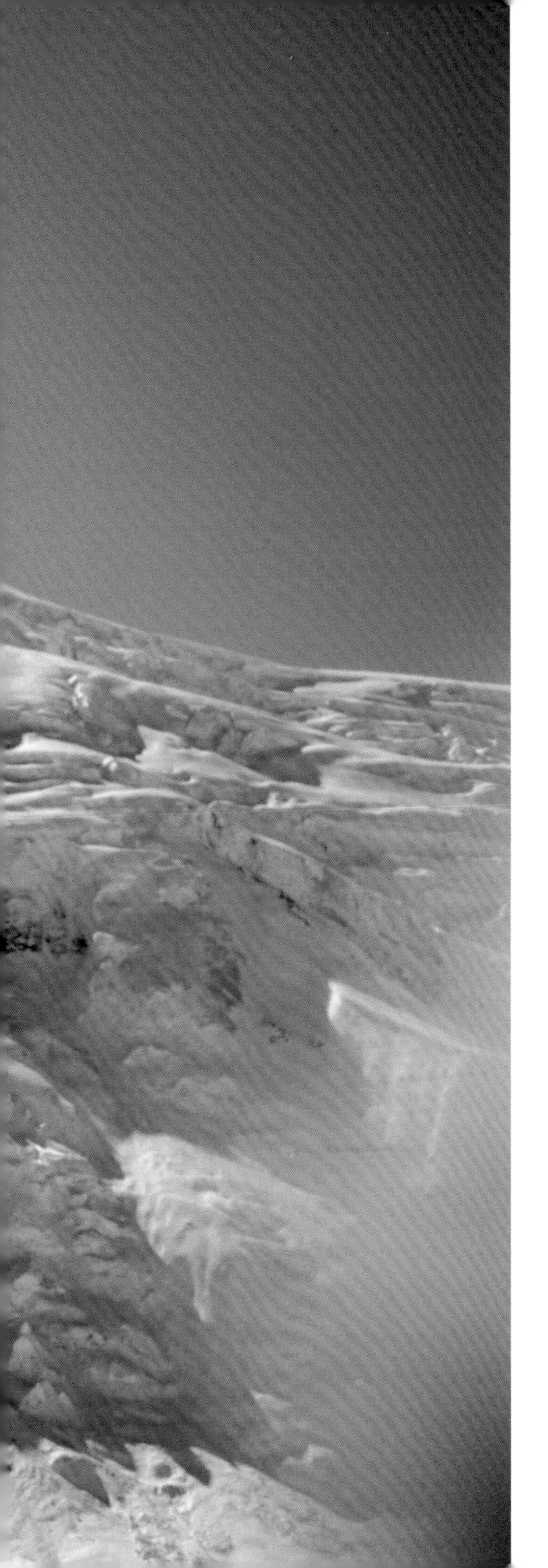

CHAPTER SEVEN

# THE THIRD POLE

*The Himalayas reach to the high-altitude jet streams that encircle Earth.* (PAM)

*Cultivated fields in the foothills of the Himalayas.* (PAM)

*One of the many "string" bridges that cross the steep valleys of the Nepalese Himalayas on the way to Mt. Everest. Though the bridges are hundreds of feet above the valley bottom, spring floods are so savage that they can launch boulders high enough to destroy them.* (PAM)

## INTO THE MOUNTAINS OF NEPAL

KATHMANDU IS THE GATEWAY TO NEPAL—the country with the greatest vertical relief in the world. Centuries of hand labor have created terraces that stretch up the valley sides, forming a lush greenway into the high mountains. I always enjoyed the walk and climb up into the mountains from the lowlands and our expedition in the fall of 1992 was no different. As we left the lower elevations behind, our pace steadily slowed, impeded by decreasing oxygen levels. At 20,000 feet there is only half the oxygen that there is at sea level. The locals, however, especially Sherpas, were as strong as ever, even at the highest elevations.

Working with the Sherpas was a humbling experience. I typically monitor the blood oxygen saturation of our team members as one indicator of our fitness and health. At sea level, oxygen saturation is usually more than 98 percent to 99 percent for a healthy, resting person. At altitude, it usually drops to the mid-80s and lower—part of the reason one is not able to work as hard at altitude. I always figured that the Sherpas, as well acclimatized as they are, must have oxygen saturation levels at altitude that were comparable to ours at sea level, since they seemed to literally skip up the mountains. But when I had the chance to regularly measure their oxygen saturation on one of our longer trips with them, it was no higher than ours.

We worked from east to west across Nepal, looking for sites that might hold a good ice core record. In 1992, we were no longer able to find suitable ice that preserved an environmental record at the 15,000-foot levels as we had 10–15 years earlier. We found we had to go closer to 20,000 feet and higher. The retreat of these glaciers due to warming posed challenges for us, and while it is likely that we will not lose all Himalayan glaciers in the near future, we will lose the Himalayan ice core climate records in the not-to-distant future, perhaps in 10–20 years as warming saturates the glaciers and destroys the original, annually layered order of the records held within the ice.

But, even more than losing the climate records, I was thinking then, and still do today, about a bigger concern: the millions of people who depend on meltwater from Himalayan glaciers to provide food and water for their families. For them, the shrinking glaciers are a matter of life and death. To add insult to injury, the rapidly melting glaciers sometimes create lakes, dammed behind poorly consolidated piles of gravel and ice pushed up by the glaciers, known as moraines. These moraines are not good dams and are prone to bursting, sending large floods down-valley capable of wiping out whole towns. Today, many local towns are devoting enormous effort to letting the water out of these dams before they burst. They are digging giant trenches by hand—time and effort they must take away from growing food. The dangers and costs imposed on these people struck me very hard and added further fuel to my determination to understand climate and bring that understanding to the public and to policymakers.

*Entrance to the Tengboche Monastery on the route to Everest from the Nepalese side of the Himalayas.* (PAM)

*Walking through the streets of Kathmandu.* (PAM)

*Yaks carrying loads into the Everest region.* (PAM)

*Namche Bazaar on the route to Everest through the Nepalese Himalayas.* (PAM)

*Morning fog engulfing the humid foothills of the Himalayas. (PAM)*

*Holy man in Kathmandu. (PAM)*

*Airstrip at Lukla, 9,000 feet above sea level, where many tourists and climbers start the Everest trekking route. (PAM)*

I FIRST TRAVELED TO CHINA IN AUGUST 1989 and have returned several times since. On the first trip, I had the opportunity to visit several portions of the Tien Shan "Celestial Mountains" as the guest of the Lanzhou Institute of Glaciology and Geocryogeology, where Chinese glaciologists were monitoring the dramatic retreat of the glaciers throughout this region. It was fascinating to follow their research and to compare their findings with what we had found on the southern slopes of the Himalayas several years earlier.

Glaciers take a while to respond to climate change, and larger ones take longer than smaller. A glacier that is not changing in size tells of a relatively unchanging climate; a shrinking glacier tells of less snowfall, more melting, or both. By looking at many glaciers across the Himalayas and Tibet, we could construct a picture of whether—and how—climate was changing over the vast Third Pole. The Chinese were seeing the same thing we had—glaciers, large and small, across the region, with only a very few exceptions, were shrinking, rapidly. This simple yet important observation was a direct result of the conclusions being drawn by other scientists measuring increasing heat content in the oceans and atmosphere. The warming we had been expecting, but that had been temporarily hidden by industrial emissions, such as sulfate that reflect incoming solar radiation, was well under way.

The 1989, Tien Shan expedition provided an excellent opportunity to start a collaboration with our Chinese colleagues that has now taken us throughout Tibet, yielded a large number of researcher and student exchange visits, and embedded in my University of Maine Climate Change Institute many of the wonderful academic customs practiced by the Chinese.

*The Chinese refer to the Tibetan Plateau as the "Third Pole". It is an apt title as it contains the largest concentration of glaciers outside of the polar regions and has many similarities to Antarctica. Like Antarctica, it is a high altitude—15,000 feet, on average—and an arid region that influences climate throughout much of the Northern Hemisphere. At least in the winter, much of it is white with snow, acting as a giant solar reflector. Here, Annapurna is visible from the shore of Lake Pokhara in Nepal.*

*Left: A small village off the main route to Everest in the Nepalese Himalayas.*

*Right: A Nepalese woman in her rice field.* (All photos: PAM)

*Inset: On one of our reconnaissance trips into the Nepalese Himalayas, my wife Lyn was able to join us. It was great to be able to show her the people and landscape I had been describing to her for years. Photo by Arun Shrestha, former University of New Hampshire graduate student, now a hydrologist in the Integrated Centre for Integrated Mountain Development, Nepal..*

*A Nepalese home offered as an inn to travelers on the way to Mt. Everest.* (PAM)

*The Tengboche Monastery on the trekking route to Mt. Everest from the Nepalese side of the Himalayas. (PAM)*

*A monastery near Lhasa, Tibet.* (PAM)

*Yak-cow hybrids plowing fields in Tibet.* (PAM)

*View of Lanzhou, central China, and the arid landscape surrounding the town, as seen from a nearby hill.* (PAM)

*Driving toward the Mongolian border.* (PAM)

## WORKING AT THE TOP OF THE WORLD—MT. EVEREST

BY THE MID-1990S, we had identified several potential ice coring sites and collected ice core records covering the last few decades from Tibet and the Himalayas. They provided extremely valuable insight into the season-to-season and year-to-year details of the Asian monsoon, as well as natural, industrial, and agricultural changes in Asia.

In 1996, my good friend Qin Dahe, then director of the Chinese Meteorological Administration, and I had the chance to meet in Cambridge, England. Together, we had planned several earlier projects in Asia and Antarctica, and now over a pub lunch, we talked about collecting a long, highly resolved Himalayan record. This was an ambitious endeavor, but a long record from the Himalayas would be very valuable and with our combined experience, we thought we could do it. Research in the early 1960s by Maynard Miller, then at Michigan State University, had shown that the south side of Mt. Everest contained a record of atomic bomb testing and we thought it might be possible to find a well-preserved environmental record at a site suitable for ice core drilling in the region—where the ice flow had not severely contorted the record, melting was minimal, and the strong winds had not blown the record away. We decided to work together to find and recover a long record from the high northern slopes of Mt. Everest.

Over the following three seasons, and together with our former students Shichang Kang and Shuigui Hou, we were able to collect several ice cores from 21,400 feet near Mt. Everest's Advanced Base Camp. The longest core was 314 feet and, with the careful analytical work of my graduate student Susan Kaspari, now an assistant professor at Central Washington University, it yielded a year-by-year climate history going back over 1,000 years. Among many other results, this long history showed that, during warmer periods, the monsoon reached farther north than during cooler periods. It reached farther north because the colder air over the Tibetan Plateau also warmed and, as a consequence, was no longer an effective barrier to the northward flow of the warm, moist air from the Indian Ocean. The implications for modern-day warming are very important and suggest that there will be at least long-sustained periods of increased precipitation over the northern Himalayas and portions of the Tibetan Plateau in coming decades. However, the moisture may come more as rain than snow, so the water storage capacity of the glaciers will likely still diminish while the water flow and likelihood of massive water discharges from the Himalayas could increase. This and follow-up research by my current graduate student Bjorn Grigholm have demonstrated dramatic increases—in step with the growth of industry—in many by-products of industrial activity, such as lead, bismuth, uranium, cesium, copper, cadmium, and sulfur. With this research, it became clear that much of the acidity we found during our early years in Asia could, in fact, be attributed to human activity.

Collaborations on cores collected throughout Asia continue with researchers from China, India, Nepal, Japan, Germany, Switzerland, Kyrgyzstan, Tajikistan, and others in the U.S., much of it part of an effort called the Asian Ice Core Array (AICA) that I lead with Vladimir Eizen from the University of Idaho. We are in the process of providing a detailed view of the rise of industrial and agricultural activity in Asia, and the much-needed, deeper understanding of how the monsoon has changed in the past under naturally warm and cold periods, allowing greatly improved predictions of water availability for the millions—and even billions—who depend on it.

In all our expeditions, friendship is perhaps the most important element. Bonds with people like Qin Dahe drive enthusiasm and adventure, and go beyond international borders to produce climate insights that might otherwise be slow in coming. My friendship with Dahe also made a big difference in evacuating five of us from a bad car accident on the northern edge of Tibet when one of us, Shichang Kang, used a satellite phone to call Dahe in Beijing who, in turn, mobilized vehicles located hours from the crash that got us to hospitals. We all recovered and the non-Chinese among us got to experience a unique combination of Western technology and Eastern pharmaceuticals that brought us back to good health.

Most driving experiences in Asia are wonderful adventures, and our accident was an exception. For example, on our last departure from Mt. Everest in 2000, we had the good fortune to hire an old Land Rover to take us from the edge of the Tibetan Plateau down the south side of the Himalayas into Kathmandu. As we steadily lost altitude, the dry, cold plateau gave way to progressively more verdant and welcoming valleys filled with rice terraces, banana plants, and innumerable flowers. It was a sensory treat I will not soon forget.

*The north side of Mt. Everest from the Rongbuk Monastery. (PAM)*

*Monks walking down from their small mountaintop monastery to fetch water from the arid Tibetan Plateau, as seen on our route to Mt. Everest through Tibet. (PAM)*

*Prayer flags on a ridge in Tibet, with a view of the Everest region. (PAM)*

*Clean suits are required to prevent contamination of the snow samples. This photo is from a snowpit in the Nepalese Himalayas near Annapurna.* (PAM)

*Traversing a glacier slope as we enter the Mustang region of the Nepalese Himalayas.* (PAM)

*The Potala Palace in Lhasa, Tibet.* (PAM)

*Left: Taking a break from carrying loads into the Annapurna region on one of our ice coring expeditions.* (PAM)

*Below: The Land Rover we hired to drive from the edge of the Tibetan Plateau down-valley into Kathmandu, Nepal.* (PAM)

CHAPTER EIGHT

# EXPLORING A CLIMATE LIBRARY

## *The International Trans Antarctic Scientific Expedition (ITASE)*

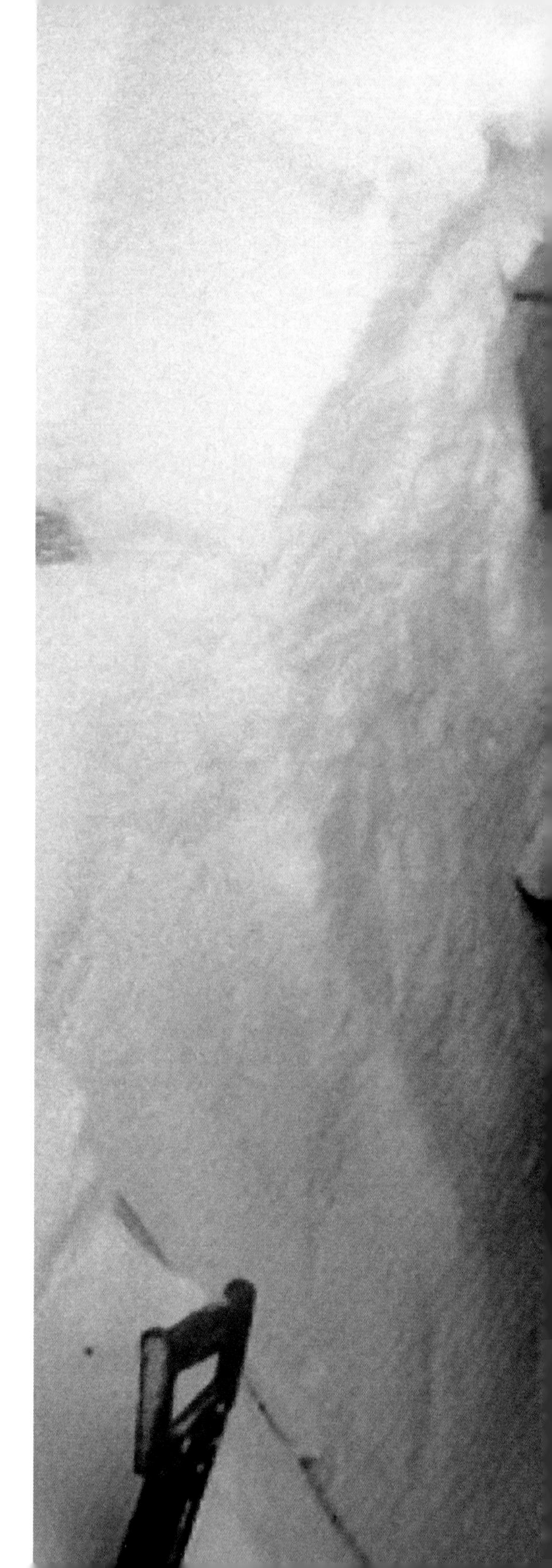

*Climate Change Institute graduate students Dan Dixon, left, (now a postdoctoral fellow in the University of Maine Climate Change Institute), and Vandy Blue Spikes, (former University of Maine graduate student now at the Earth Science Agency LLC), sampling in a snowpit on the East Antarctic Pleateau on a windy day.* (PAM)

## CROSSING ANTARCTICA

SEVERAL ICE CORE CLIMATE RECORDS have been collected in Antarctica, but Antarctica is a big continent, and as of the late 1980s, the records that had been collected revealed only a small portion of its vast and complex climate picture. However, a broader, more detailed set of records was essential to understanding critical features of Antarctic climate change—a place that is home to enough ice to raise sea level by more than 200 feet, and influences climate worldwide, including currents in the ocean and atmosphere, the amount of sunlight absorbed or reflected by ice and snow on land and the ocean, and biological activity throughout the enormous Southern Ocean.

In 1989, at an international meeting in Grenoble, France, I proposed that Antarctic researchers take on the goal of developing such a record under the name International Trans Antarctic Scientific Expedition (ITASE). I suggested a series of over-snow traverses to collect ice cores and snow samples, survey the ice thickness, measure the ice sheet's flow direction and speed, install automatic weather stations, and more. Preparing for the meeting in our hotel room, my wife Lyn put her professional artistic talents to work and we sketched a diagram showing the proposed effort. The idea was well received, and ITASE came into being. Today, 21 countries participate.

As field leader of the U.S. portion of the ITASE over-snow traverses, I led all of the U.S. ITASE expeditions (1999–2003; 2006–08) logging 16 months in the field and more than 6,000 miles over unexplored sections of the ice sheet. In addition to the data we collected, our experiences in the field provided on-the-ground insight into Antarctica's diverse and dramatic climate—insight that has been invaluable in making sense of the data from our ice cores, weather stations, and other instrumental records. We encountered storms that nearly buried our vehicles, terrain that was so soft we could not easily move forward without sinking, and rock-hard, wind-polished ice surfaces that shattered as the vehicles passed over, and more.

Though Antarctica is extremely remote, and has often been thought of as isolated from human activity, then graduate students Dan Dixon and Elena Korotkikh from our Climate Change Institute developed records that have revealed high levels of lead, uranium, cadmium, copper, and mercury—all rising with industrial development, just as we have seen in other ice cores from around the globe.

Antarctica's vast store of ice is one of the best repositories of historical climate available and for periods before instrumental records, ITASE efforts have turned Antarctica from one of the least-understood portions of the Earth's climate system into one of the best understood. Today, we realize that Antarctic climate is changing quite notably in response to greenhouse gas rise and ozone depletion, and that changes in the extent of the Antarctic Ice Sheet, its glaciers, and sea ice all exert major impacts on the climate over the Southern Ocean, Australia, New Zealand, South America, and Africa.

*Mark Wumkes fixing the ice coring drill, with insulated boxes containing ice cores in the background. Despite temperatures many degrees below freezing, white insulated boxes are needed to keep cores from melting in direct sunlight.* (PAM)

*Photograph taken from the roof of a sled mounted with solar panels. We stopped every six miles at GPS waypoints to collect surface snow samples for chemical analyses, to check if equipment was properly lashed down, and to give team members a chance to stretch their legs, since we usually travelled less than two to three miles per hour. Photo by Dan Dixon, Climate Change Institute, University of Maine.*

*View of a crevasse detector, as seen from the driver's seat of the lead tractor. The radar system can detect crevasses about 60 feet ahead—ideally enough time to stop and choose a different route.* (PAM)

*Small snow dunes, known as sastrugi, forming in the lee of our vehicles after just a few hours of blowing snow. Photo by Dan Dixon, Climate Change Institute, University of Maine.*

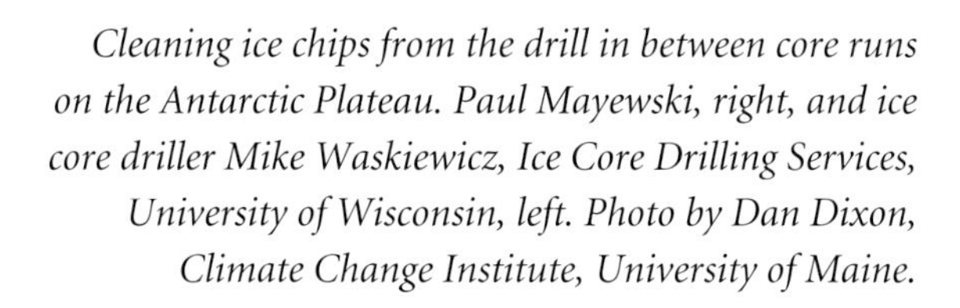

*Cleaning ice chips from the drill in between core runs on the Antarctic Plateau. Paul Mayewski, right, and ice core driller Mike Waskiewicz, Ice Core Drilling Services, University of Wisconsin, left. Photo by Dan Dixon, Climate Change Institute, University of Maine.*

## EARTH'S SLEEPING GIANTS

IN 2009, SEVERAL OF US (John Turner, Robert Bindschadler, Pete Convey, Guido di Prisco, Eberhard Fahrbach, Julian Gutt, Dominic Hodgson, Paul Mayewski, Colin Summerhayes) edited, and with many others contributed to, a synthesis report for the Scientific Committee for Antarctic Research (SCAR) titled, "Antarctic Climate Change and the Environment" (ACCE). Our synthesis highlighted the fact that some of the greatest warming on the planet is occurring around the Antarctic Peninsula, as well as the immense potential for future change in a warming world. Among other effects, the warming thus far has caused the dramatic, rapid disintegration of enormous ice shelves—floating tails of glaciers extending out into the ocean. Ice cover in coastal West Antarctica is anchored to a few islands of land and it, too, is particularly vulnerable to the type of disintegration we have already seen in the Antarctic Peninsula. Projected melting based on a middle-of-the-road model estimate of 5.4 °F rise by 2100 would incur a high cost in sea level rise, with the current estimate from our report being up to 4.6 feet by 2100. We have come to describe the recent condition of the Antarctic and Greenland ice sheets as Earth's "sleeping giants." Though safe enough while sleeping, the loss of some of the Antarctic Peninsula's ice shelves, like many other effects of warming, are signs that these dangerous giants are waking.

*Extensive ice loss on land and sea is occurring throughout the Antarctic Peninsula, but intensification and shrinkage of the belt of Southern Hemisphere westerly winds is insulating the rest of the continent from warm, northern air masses. This recent change in the behavior of the westerlies is a product of the human activities that created the Antarctic ozone hole and the rise in greenhouse gases in the lower atmosphere.* (PAM)

## SURPRISE AND INSTABILITY

ACID PRECIPITATION caused by fossil-fuel combustion, especially coal-fired power plants, has had devastating effects on forests, waterways, ecosystems, buildings, and human health. It was a tremendous political achievement that we were able to dramatically reduce these pollutants in the late 20th century. The CFC-caused ozone destruction was also poised to have catastrophic effects, and we should give ourselves a great deal of collective credit for an effective response. But with Earth's climate, all things are connected. Until we cleaned up the sulfate we had been putting in the air, we didn't understand that its sunlight-reflecting character had counteracted some of the warming effect of carbon dioxide from the 1940s to the 1970s.

We are lucky that the oceans have absorbed about half of the carbon dioxide produced by industrial activity. Otherwise, atmospheric concentrations would now be closer to 500 parts per million instead of our current 390. But, the absorbed carbon dioxide is acidifying the oceans—acidification that is close to the point where it will dissolve, like chalk in vinegar, the calcium carbonate shells of the microorganisms that are the primary source of food for most marine life and that produce a large portion of the oxygen in the air we breathe.

As my graduate student Dan Dixon and I studied the results of our ITASE ice cores, we saw that the combination of greenhouse gas heating and ozone destruction worked together to intensify and contract the westerly winds that circle Antarctica, as demonstrated earlier by David Thompson (Colorado State University) and Susan Solomon (National Center for Atmospheric Research). With the ice cores, we could provide a longer perspective showing that the current intensification is unique in the last few hundred years—yet another demonstration of the human imprint on climate.

While Antarctica is losing ice in coastal regions as greenhouse heating continues, the increased speed and poleward migration of the westerly winds have resulted in the temporary isolation and protection of Antarctica's interior from the now warmer, northern air. Remarkably, the intensification and contraction of the westerly winds has also helped to maintain coastal sea ice around much of the continent by blowing it landward rather than seaward into the warming ocean. Alternately, because the Antarctic Peninsula sticks out significantly north of the main Antarctic continent, the combination of Southern Ocean warming and increased westerlies leads to melting of sea ice and blowing it offshore.

In coming decades, as ozone-destroying CFCs decrease in the atmosphere, high-altitude ozone recovers, and greenhouse gas warming continues, we may see a return to weaker westerlies that means weakening of the barrier protecting the interior of the Antarctic from warming, resulting in a potentially dramatic increase in melting—"catching up", as it were, to where we might have been otherwise, in much the same way as global warming "caught up" as we reduced our emissions of sulfate.

Examination of our long Antarctic ice core records suggests that the last time the westerlies intensified was more than 5,000 years ago, likely a consequence of natural changes in high-altitude ozone caused by changes in the Sun's output that affected ozone levels. In the past, westerly intensifications were short-lived and when they ended, they did so abruptly in a matter of a few years. Now, when human activity has resulted in ozone depletion and subsequent healing, it is certainly possible that the westerlies will weaken, rapidly breaking down the protective barrier, resulting in a rapid increase in surface melting in the interior regions and associated consequences. It was essential that we reduce ozone depletion, just as it was essential that we reduce sulfate levels in the Northern Hemisphere, but in both cases they temporarily masked other human impacts.

Humanity clearly alters climate in many and significant ways, but until recently it has done so without understanding the consequences—or even the knowledge that we were altering it. The changes have never been without surprises, and have always taken us away from the stable climate in which Civilization arose and toward an unknown and unstable future.

*Aerial view of Tasman Glacier on the South Island of New Zealand. In response to strong warming in the Southern Hemisphere, this glacier is melting rapidly and forming the massive meltwater lake seen here.* (PAM)

CHAPTER NINE

# BEFORE THEY DISAPPEAR

*Iceberg and penguin drifting north from the Antarctic Peninsula.* (PAM)

## VOLATILITY ON THE ANTARCTIC PENINSULA

AS THE ITASE TRAVERSES PROGRESSED, we also expanded north from the main Antarctic continent in our recovery and interpretation of ice cores, into the Antarctic Peninsula, South America, and New Zealand. It is clear from instrumental records and from ice and lake core records, that changes in Antarctic climate have impacts on the Southern Hemisphere and global climate. The populated regions impacted by changes in Antarctic climate are the sub-Antarctic islands—Australia, New Zealand, southern South America, and Africa. The need to document and understand changes in these regions is essential, but just as in the Himalayas, collection of unaltered, environmentally relevant ice core records is being affected by greenhouse warming. In spite of the immense heat-holding capacity of the Southern Ocean, glaciers north of Antarctica's temporary protective wind barrier are seeing the effects of warming.

On visits to the Antarctica Peninsula and other sub-Antarctic islands in the 1970s, we found scientific stations with easy access to other islands and the mainland via permanent sea ice bridges. On more recent visits in 2008, we found that these stations had become isolated outposts on craggy islands, entirely dependent on ship and aircraft. At one station on King George Island, I spoke with U.S. researchers Wayne and Susan Trivelpiece, who described an 80 percent decrease in krill—and a subsequent decline in penguin populations that depend on krill for food—as a result of sea ice loss.

Work conducted in 2007 and 2008 by colleagues Andrei Kurbatov and Mario Potocki from our Climate Change Institute, Jefferson Simoes from Brazil, and Ricardo Jana from Chile have also revealed the effects of industrialization in the Southern Hemisphere. Their ice core records show recent, dramatic increases of copper, cadmium, and uranium from the rapidly growing open-pit mining activities in Australia, South America, and Africa.

*The Antarctic Peninsula is the region of Antarctica that is most vulnerable to greenhouse warming. Warming has melted enormous areas of sea ice, exposing the darker, less reflective, heat-absorbing ocean surface. Sea ice loss further warms the region as this open water allows ocean heat to be released to the atmosphere. (PAM)*

## THE WIND NEVER STOPPED IN TIERRA DEL FUEGO

IT WAS LATE EVENING IN MARCH 2006 and we were standing on the edge of a pier in Punta Arenas, Chile, waiting for a small wooden fishing boat to take us on a 12-hour sail to a beach below the Cordillera Darwin in Tierra del Fuego, our intended ice core drilling site. The man standing next to me asked in English what we were doing and I explained that we were on an expedition to one of the southernmost ice caps in South America to recover a record of the climate history in the region. I mentioned what an amazing day it had been and that on previous trips to the region, I had never been able to see as far as I could that day. He looked over and said, "Ah yes. Today we have a Sarmiento High," named after the distant mountain visible on days like this. In the same breath he added, "but it won't be so nice in a few hours." The import of his prediction was soon to unfold.

We left on a small wooden fishing boat that night, had an absolutely placid crossing of the famous Magallanes Strait, and landed sufficient stores for a two-month stay, because we knew from Chilean colleagues how treacherous the weather could be and understood there was a possibility we might be stranded for many weeks. Right on schedule, the helicopter I chartered arrived at the beach to lift the four of us and gear to the Cordillera Darwin ice plateau. I had always climbed into the mountains in the past, but this ice plateau was surrounded by very nasty looking crevasses and, for once, we were in a mountainous region with helicopter availability.

I was on the first of what were to be three flights. The site was stunningly beautiful, ringed with mountains and filled with a clear light that brought out minute details. By the second and third helicopter lifts, a mere 30–60 minutes later, I began to realize there might be some truth to what the man on the Punta Arenas dock had said. The clouds came in and it began to storm. Seventeen days later, we had spent almost all of our time just shoveling out our camp from the continual heavy snows and furious winds that often reached over 100 knots. On the seventeenth day, the helicopter reappeared. The pilot, Marcello Lira, told us he had been forcibly thrown into our camp by a 110-knot wind that opposed his 115-knot forward air speed. The swirling winds tossed his helicopter out of control twice while attempting

to land, and he appeared at our site almost surprised—his intention having been to go back to the beach to wait, again, for the winds to calm. The four of us jumped in with what we could carry and were thrown back into the sky by the wind. Safely on the beach, there was a good deal of hugging and heartfelt thanks to Marcello. Then we waved goodbye to Marcello and waited for an even smaller boat than the first fishing vessel to bring us back to Punta Arenas on a very much less placid crossing.

Ice core records from this critical region are essential to understanding future climate change. For example, between Tierra del Fuego and the Antarctic Peninsula, the Drake Passage plays a critical role in the climate system. Here, westerly winds are forced through their narrowest constriction rounding the Southern Hemisphere, and small northerly or southerly shifts of the westerly winds can have significant climate impacts. These are the very shifts that might provide insight into the future of drought in Australia, glacier melting and subsequent water availability in South America and New Zealand, the impacts of future warming, and other fast-moving changes.

*Left: The Cordillera Darwin Range in Tierra del Fuego as it appears during the infrequent and short-lived, "Sarmiento highs." Photo by Andrei Kurbatov, Climate Change Institute, University of Maine.*

*Andrei Kurbatov (Climate Change Institute) prepares equipment during a brief moment of calm in a 17-day storm on the Cordillera Darwin Range, Tierra del Fuego.* (PAM)

*The helicopter departs our Cordillera Darwin drilling site for another load from the beach during camp set up. The clouds turned into a serious storm before the helicopter returned minutes later.* (PAM)

*Examining damage and picking up the pieces of a tent that was destroyed during a major wind gust. Andrei Kurbatov and Bjorn Grigholm, left and right front, respectively, Paul Mayewski in background. Photo by Mark Wumkes.*

*Back down on the beach near the Cordillera Darwin Range after the 17-day storm. Left to right: Bjorn Grigholm, Paul Mayewski, Marcello Lia, Mark Wumkes and Andrei Kurbatov. Photo by Alejo Contraras, Aerovías DAP, Las Aerolíneas de la Patagonia.*

*The fishing boat that brought us from Punta Arenas to the beach below the Cordillera Darwin Range, Chile.* (PAM)

*Andrei Kurbatov collecting wood for a campfire on the beach near the Cordillera Darwin as we wait for a fishing boat to take us back to Punta Arenas.* (PAM)

## BEFORE NEW ZEALAND'S GLACIERS MELT

ALTHOUGH A RELATIVELY SMALL COUNTRY, New Zealand offers an amazing range of landscapes—from rain forests to volcanic peaks, deserts, and glaciers. Set in the center of the South Island of New Zealand are several glaciers that might provide robust ice core climate records. In collaboration with New Zealand colleague Uwe Morgenstern, we have been collecting ice cores from these glaciers since 2004 and are now examining the records. Our hope is to recover a several hundred-year record of temperature, precipitation, and storm patterns, and understand the changes in atmospheric chemistry that mark the transition from native Maori culture to colonization by Europeans, and the onset of modern farming practices and industrial activity. Like the Himalayas, as New Zealand glaciers retreat, many towns and farming communities will lose the year-round water supply on which they depend.

Our ice core expeditions in the New Zealand Alps are a very different experience from many of our usual remote experiences. Travel to the field site most often begins in the beautiful English city of Christchurch on New Zealand's South Island, and from there by car through the grazing pastures of the east coast to Mt. Cook tourist village—a comfortable and scenic journey. Transport to the field is also a simpler undertaking than in many other locations—a few minutes' flight in a Pilatus Porter, a short takeoff and landing single-engine aircraft, brings us to a ski-landing close to our chosen study site. However, this is still serious terrain. The New Zealand Alps are formidable mountains that require attention and humility, and we have often been pinned down by strong winds and blowing snow for several days.

From our work in New Zealand, we hope to track changes in the westerly winds—changes that are a result of the combination of the Antarctic ozone hole and greenhouse gas warming. We also hope to follow the regional history of industrialization and contribute to making better predictions of New Zealand climate.

In the New Zealand Alps, as with the South American Andes and the Himalayas, all the high-altitude glaciers that might hold good environmental records are in retreat and in danger of melting, to the extent that they will no longer produce intelligible records. The impacts of greenhouse heating mean that the generation of glaciologists trained today might be the last who will have the opportunity to collect such records.

*Left: Sunset view from our Tasman Glacier ice drilling camp in the New Zealand Alps. (PAM)*

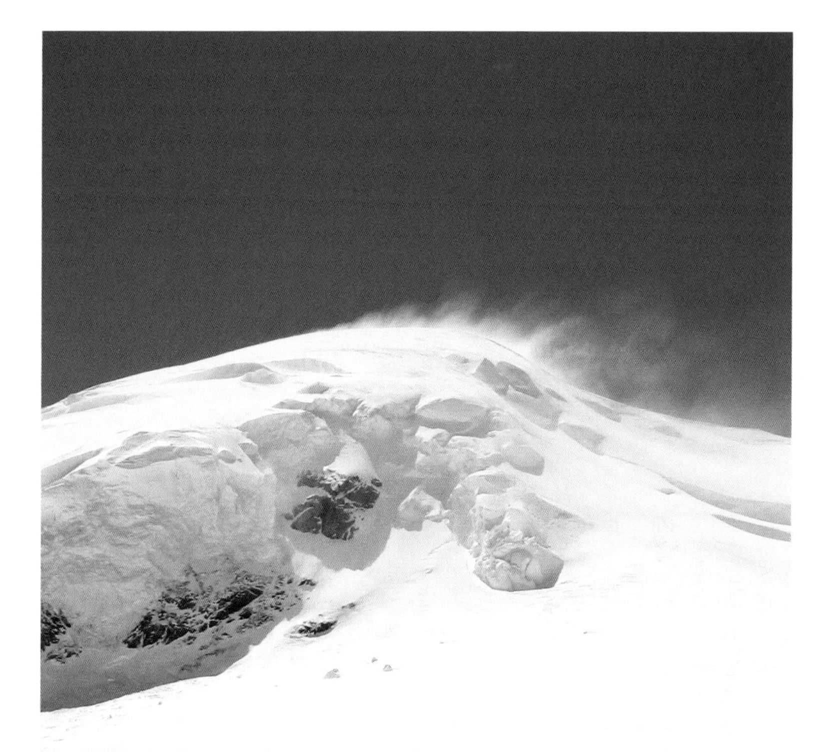

*Uwe Morgenstern (GNS, Wellington, New Zealand) helping set up our ice coring drill on the Tasman Glacier in the New Zealand Alps. (PAM)*

## A VOLCANO ON THE ANDEAN SPINE

ON FEBRUARY 27, 2010, our tent walls appeared as if they themselves were luminescent in the brilliant, full moonlight. We had successfully collected a 30-foot reconnaissance ice core from the Tupungatito Glacier at about 18,700 feet and were now in the thick air of our lower camp before the trek out to the mining road. I (Michael Morrison) had joined a team led by Paul and Chilean colleague Gino Casassa, along with several Climate Change Institute members, a Chilean logistics manager Gonzalo Campos, four Chilean arrieros, 10 mules, and four horses.

At about 3:40 a.m., the first waves shook us forward and then back, waking us from our sound, oxygen-rich sleep. "An earthquake," I thought as I wondered how big it was—either larger and farther or smaller and closer. In the next instant, I mentally reviewed our camp's location and risk of rockfall from the tremors. Our camp, by choice, was in an area safe from rockfalls, and we were grateful for our attention to detail in picking the site.

We all came out from our tents as the sound of thousands of large and small rockfalls throughout the valley rang in our ears. Within 15 minutes, the full moon was obscured by the rising dust. Though some built-up areas of Chile were damaged and, sadly, lives were lost in the earthquake and ensuing tsunami, our experience in the mountains was one of fascination in witnessing this rare event. However, earthquakes were not the reason for our expedition, and rockfalls may not be the only cause of hazy air in the high, Andean valleys.

The surface waters of the Southern Pacific warm and cool every few years, with warm years known as El Niños and cold as La Niñas driven by oscillations in atmospheric pressure across the South Pacific. Tupungatito, we hoped, would give us an up-close view of these warming and cooling surface waters, as well as the important Antarctic westerlies to the south. Like the other "boundary" regions we had studied, these temperature swings would be preserved as changes in the chemistry of the snow falling on the glacier, and in the amount of snow that falls in any given year.

There are not many suitable glaciers in the Andes for recovering ice core records. Many slide down steep mountains too fast, others melt too much, and still others are politically challenging to access. Almost all of them are retreating, and finding appropriate glaciers is a race against time. Along with Gino Casassa, we were lucky to find the Tupungatito Glacier. It is deep, relatively flat, and appears to be still cold enough to contain a good history. We will go back and drill a core to the bottom as soon as possible.

We anticipate that it will also tell us about the persistent haze in the air, even at high altitude and before the earthquake—haze that may come from a combination of the arid landscape and the copper smelters that have been built in recent decades. As Chile and other South American nations build up their industrial infrastructure, they too are beginning to face the same issues we face in the Northern Hemisphere and, if our coring site is as good as it promises to be, it will tell the climate story of this region.

*Looking across one of the Tupungatito craters toward the Tupungatito Glacier where we drilled an ice core. (MCM)*

*Arrieros and mules crossing the Colorado River on the way to the Tupungatito Glacier in the Chilean Andes.* (MCM)

*Horses and mules graze on the grasses at 10,000 feet above sea level on the route to Tupungatito Glacier in the Chilean Andes. (MCM)*

*Aconcagua to the north from a high camp at about 16,000 feet. In the foreground are slivers of snow left from winter, known as penitentes, formed through sublimation and melting. (MCM)*

*Returning from the drill site on the glacier to our camp at 16,000 feet. (MCM)*

*An acclimatization camp at 14,000 feet. (MCM)*

CHAPTER TEN

# UNDENIABLE

WITHOUT PERSPECTIVE, it would be hard to demonstrate change either from place to place or from time to time. Hence the journey in search of such perspective that forms the substance of our work and this book. The figure on this page is a synopsis of some, though not all, of the important air quality findings that we have discovered on this journey. It provides undeniable examples of human contributions to the atmosphere—undeniable because there can be no rational debate that the changes portrayed are real and the source for the overwhelming majority of them is human activity.

The figure is divided into two parts. The section on the left covers the last 5,000 years and the right from 1700 AD to 2000 AD. We start with the rise in population, since it is not only the emissions related to human activity that are important, but also the number of humans emitting these substances. The remainder of all of the information in the figure comes from ice core records, except the CFC-11 record. For the two greenhouse gases in the figure, carbon dioxide and methane, the most recent 50 years of each come from observations conducted at the Mauna Loa Observatory, Hawaii. For some measurements, the record does not yet extend back more than 200–300 years before present, but it will with future work. A full range of pollutants that would show similar, dramatic human increases would include the toxic metals uranium, cesium, cadmium, and copper; volatile organic compounds; dust and other particulate matter; ground-level ozone; and many more humanly engineered chemicals. For almost all of these records, the unique and rapid observed rise has been demonstrated by ice core records to be of human origin.

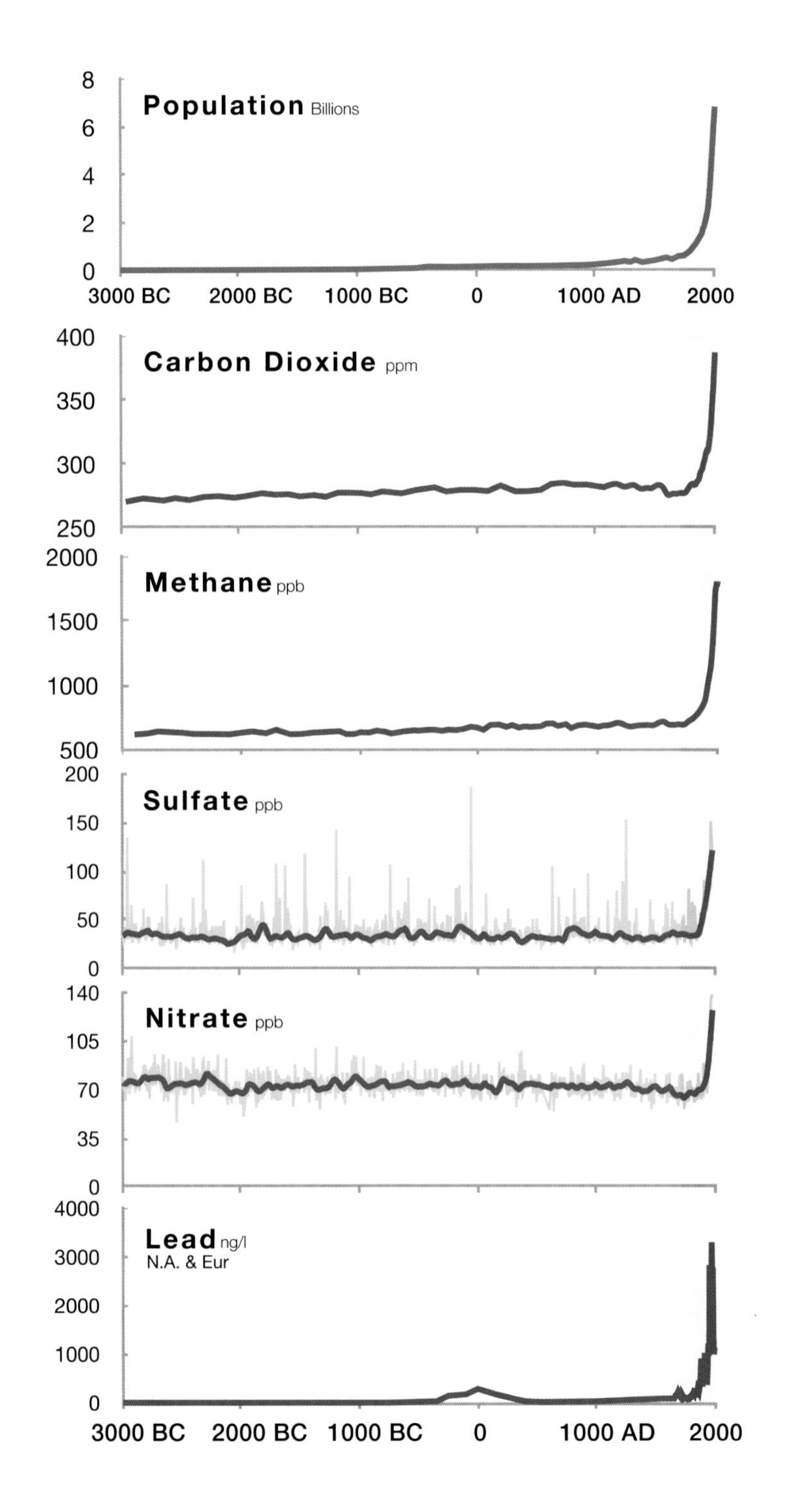


*Ice core records provide a framework for assessing human impacts on the chemistry of the atmosphere. At right, examples for the last 5,000 years; and at the far right, the last 300 years. See text following figures for details and references.*

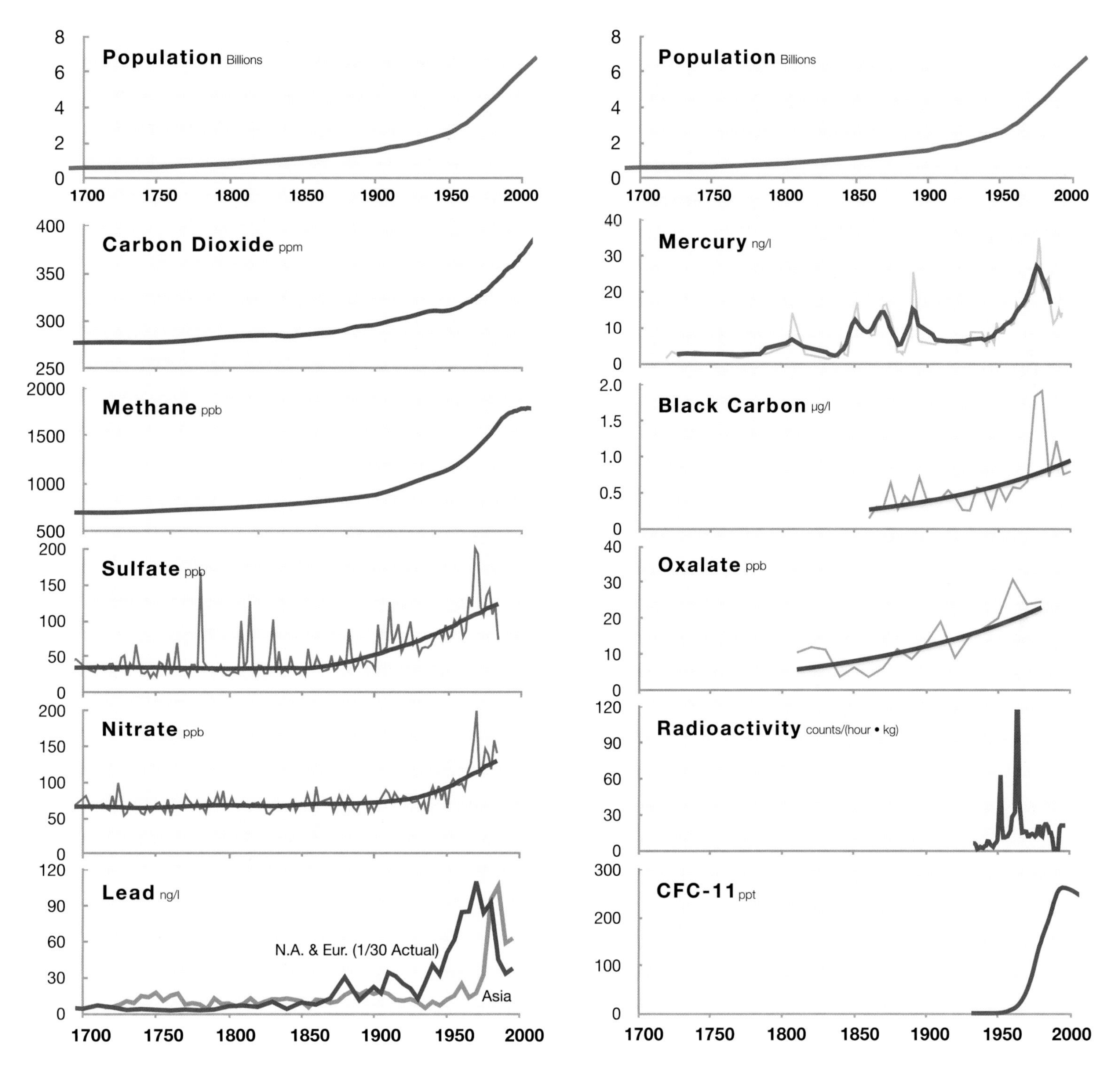
Population Billions
Population Billions
Carbon Dioxide ppm
Mercury ng/l
Methane ppb
Black Carbon µg/l
Sulfate ppb
Oxalate ppb
Nitrate ppb
Radioactivity counts/(hour • kg)
Lead ng/l
N.A. & Eur. (1/30 Actual)
Asia
CFC-11 ppt

The common denominator for all of these is that, at their current elevated concentrations, they are absolutely or potentially hazardous to human and ecosystem health, and all have risen significantly since the turn of the 19th century. For several, including the greenhouse gases—sulfate, nitrate, black carbon, and ozone-destroying CFCs—the impacts on Earth's heat balance and hydrology are critical. For carbon dioxide and methane (and nitrous oxide, not shown here)—all important greenhouse gases—Antarctic ice cores have shown that the modern rise is unparalleled in at least the last 850,000 years. It may turn out that the rise in methane began with the rise of intensive agricultural activity several thousand years ago. Twentieth- and 21ST-century levels of sulfate and nitrate, the precursors for acid rain, are far higher than they have been during the last 12,000 years—the period during which Civilization emerged. Sulfate is one of the few compounds that has, at times, been higher than today. After a large volcanic eruption, sulfate levels can reach very high concentrations in the atmosphere. However, unlike the sustained emissions from human activity resulting in persistently high concentrations, volcanic events are short-lived, with high concentrations lasting a year or two at most before returning to low, "background" levels.

The last years of sulfate and nitrate emissions reveal the ups and downs of industrial and economic activity, and the rise of automotive emissions. Most importantly, it also shows the leveling off and decline of sulfate since the Clean Air Act and its subsequent strengthening, demonstrating the profound success of our policy response.

Unlike carbon dioxide and other long-lived greenhouse gases that mix almost uniformly throughout the atmosphere, others of these components are short-lived and have a more regional or hemispheric impact. The effects of sulfate and nitrate have been strongest in the Northern Hemisphere, where they have historically been emitted. We are now seeing the rise of lead, uranium, copper, sulfate, and other pollutants in the Southern Hemisphere, just as they did in Northern Hemisphere a few decades earlier, in step with the industrial development now under way. We have seen atmospheric lead levels from the North American and European use of leaded gasoline begin their rise in the 1940s and decline sharply after the ban of this additive. As the use of leaded gasoline in Asian countries picked up a few decades later than in North America and Europe, ice cores downwind from Asia show a later peak and policy-driven decline. Note, in particular, that even the mining and smelting activities of the Greek and Roman civilizations are represented by a peak in emission levels about 2,000 years ago.

Climate is more than just temperature. Greenhouse warming is of utmost concern and has serious consequences for health, including vector-borne diseases, drought, flooding, and storms, but greenhouse gas rise is not the only concern. Ice cores have revealed that in addition to warming, human industrial activity has deeply affected the environment in which we live. The pollutants we have introduced to the air, water, and soil have significant health impacts for humans and ecosystems, and impose large economic and quality-of-life costs. Our policy responses to date, most notably the Clean Air Act and the Montreal Protocols, have dramatically relieved us of some of these health and economic burdens. Who would not advocate living in a healthier environment, especially when it comes with strong economic benefits? Even without economic benefits, who could look into the eyes of their children and grandchildren and say, "I am not willing to support the effective policies that will leave a cleaner, healthier, stronger world for you?"

REFERENCES FOR CHAPTER 10

■ POPULATION – source: U.S. Census Bureau (raw numbers).

■ CARBON DIOXIDE – sources: Etheridge, D.M., Steele, L.P., Langenfelds, R.L., Francey, R., Barnola, J.M., and Morgan, V.I., 1996, "Natural and anthropogenic changes in atmospheric $CO_2$ over the last 1000 years from air in Antarctic ice and firn," *Journal of Geophysical Research* 101, 4115-4128. (Law Dome region, East Antarctica, ice core, raw data); Indermuhle, A., Stocker, T.F., Joos, F., Fischer, H., Smith, H.J., Whalen, M., Deck, B., Mastroianni, D., Tschumi, J., Blunier, T., Meyer, R., and Stauffer, B., 1999, "Holocene carbon-cycle dynamics based on $CO_2$ trapped in ice at Taylor Dome, Antarctica," *Nature* 398, 121-126. (Taylor Dome, East Antarctica, ice core, raw data), Data from March 1958 through April 1974, from C.D. Keeling of the Scripps Institution of Oceanography. Data from 1974 through the present from P. Tans, NOAA/ESRL(www.esrl.noaa.gov/gmd/ccgg/trends/). (one-year average).

■ METHANE – sources: Etheridge, D.M., Pearman, G.I., and Fraser, P.J., 1992, "Changes in tropospheric methane between 1841 and 1978 from a high accumulation rate Antarctic ice core", *Tellus* 44B, 282-294. (Law Dome region, East Antarctica, ice core, raw data); Blunier, T., Chappellaz, J., Schwander, J., Stauffer, B., and Raynaud, D., 1995, "Variations in atmospheric methane concentration during the Holocene epoch," *Nature*, 374, 46-49; Blunier, T., Chappellaz, J., Schwander, J., Dällenbach, A., Stauffer, B., Stocker, T., Raynaud, D., Jouzel, J., Clausen, H.B., Hammer, C.U., and Johnsen, S.J., 1998, "Asynchrony of Antarctica and Greenland climate during the last glacial," *Nature*, 394, 739-743 (Greenland and Antarctic ice core, raw data); Dlugokencky, E.J., Steele, L.P., Lang, P.M., and Masarie, K.A., 1995, "Atmospheric methane at Mauna Loa and Barrow observatories: Presentation and Analysis of In-Situ Measurements," *Journal of Geophysical Research* 100, 23,103-113. (one-year average).

■ SULFATE – sources: Mayewski, P.A., Lyons, W.B., Spencer, M.J., Twickler, M.S., Koci, B., Dansgaard, W., Davidson, C., and Honrath, R., 1986, "Sulfate and nitrate concentrations from a South Greenland ice core," *Science* 232: 975-977; Mayewski, P.A., Lyons, W.B., Spencer, M.J., Twickler, M.S., Buck, C.F., and Whitlow, S., 1990, "An ice core record of atmospheric response to anthropogenic sulphate and nitrate," *Nature* 346(6284): 554-556; Mayewski, P.A., Meeker, L.D., Twickler, M.S., Whitlow, S.I., Yang, Q., Lyons, W.B., and Prentice, M., 1997, "Major features and forcing of high latitude northern hemisphere atmospheric circulation over the last 110,000 years," *Journal of Geophysical Research* 102, C 12, 26,345-26,366. (Greenland ice cores, five-year average for longer record, two-year average for shorter record).

■ NITRATE – Mayewski, P.A., Lyons, W.B., Spencer, M.J., Twickler, M.S., Koci, B., Dansgaard, W., Davidson, C., and Honrath, R., 1986, "Sulfate and nitrate concentrations from a South Greenland ice core," *Science* 232: 975-977; Mayewski, P.A., Lyons, W.B., Spencer, M.J., Twickler, M.S., Buck, C.F., and Whitlow, S., 1990, "An ice core record of atmospheric response to anthropogenic sulphate and nitrate," *Nature* 346(6284): 554-556; Mayewski, P.A., Meeker, L.D., Twickler, M.S., Whitlow, S.I., Yang, Q., Lyons, W.B., and Prentice, M., 1997, "Major features and forcing of high latitude northern hemisphere atmospheric circulation over the last 110,000 years," *Journal of Geophysical Research* 102, C12, 26,345-26,366. (Greenland ice cores, five-year average for longer record, two-year average for shorter record).

■ LEAD – sources: Hong, S.M., Candelone, J.P., Patterson, C.C., and Boutron, C.F., 1994, "Greenland ice evidence of hemispheric lead pollution 2 millennia ago by Greek and Roman civilizations," *Science* 265, 1841-1843. (Antarctic ice core for long record, raw data smoothed); Schwikowski, M., Barbante, C., Doering, T., Gaeggeler, H.W., Boutron, C., Schotterer, U., Tobler, L., Van De Velde, K.V., Ferrari, C., Rosman, K., and Cescon, P., 2004, "Post-17th century changes of European lead emissions in high altitude alpine snow and ice," *Environmental Science and Technology* 38, 957-964. (Colle Gnifetti ice core, Italy-Switzeerland for short record, five-year average); Osterberg, E.C., Mayewski, P.A., Fisher, D.A., Kreutz, K.L., Handley, M.J., Sneed, S.B., Zdanowicz, C., and Zheng, J., 2008, "Asian Pb emissions in the North American free troposphere from the 13,000 year Mt. Logan ice core," *Geophysical Research Letters* 35, L05810, doi: 10.1029/2007GL032680. (Mt. Logan, Yukon Territory, Canada ice core for short record, five-year average).

■ MERCURY – source: Schuster, P.F., Krabbenhoft, D.P., Naftz, D.L., Cecil, L.D., Olson, M.L., Dewild, J.F., Susong, D.D., Green, J.R., and Abbott, M.L., 2002, "Atmospheric mercury deposition during the last 270 years: A glacial ice core record of natural and anthropogenic sources," *Environmental Science and Technology* 36, 2303-2310. (Fremont Glacier ice core, Wyoming, raw plus two-year average).

■ BLACK CARBON – source: Kaspari, S., Schwikowski, M., Gysel, M., Flanner, M.G., Kang, S., and Mayewski, P.A., in review, "Recent increase in black carbon concentrations from a Mt. Everest ice core spanning 1860-2000 AD." (Mt. Everest ice core, five-year average).

■ OXALATE – source: Kang, S., Qin, D., Mayewski, P.A., and Wake, C., 2001, "Recent 180 years of oxalate (C204-2) recovered from a Mt. Everest ice core and environmental implications," *Journal of Glaciology* 47, 155-156. (Mt. Everest ice core, decadal smooth).

■ RADIOACTIVITY – source: Kang, S., Mayewski, P.A., Qin, D., Yan, Y., Hou, S., Zhang, D., Ren, J., and Kreutz, K., 2002, "Glaciochemical records from a Mt. Everest ice core: Relationship to atmospheric circulation over Asia," *Atmospheric Environment* 36, 3351-3361. (Mt. Everest ice core, one-year average).

■ CFC-11 – source: Walker Prinn, R.G., Weiss, R.F., Fraser, P.J., Simmonds, P.G., Cunnold, D.M., Alyea, F.N., O'Doherty, S., Salameh, P., Miller, B.R., Huang, J., Wang, R.H.J., Hartley, D.E., Harth, C., Steele., L.P., Sturrock, G., Midgley, P.M., and McCulloch, A., "A history of chemically and radiatively important gases in air deduced from ALE/GAGE/AGAGE," *Journal of Geophysical Research*, 105, 17,751-17,792, 2000. Walker, S.J., Weiss, R.F., and Salameh, P.K., "Reconstructed histories of the annual mean atmospheric mole fractions for the halocarbons CFC-11, CFC-12, CFC-113 and carbon tetrachloride," *Journal of Geophysical Research*, 105, 14,285-14,296, 2000. (https://bluemoon.ucsd.edu/pub/cfchist, raw).

CHAPTER ELEVEN

# THE JOURNEY

*A fjord in Tierra del Fuego, with the Cordillera Darwin Range in the background. Photo by Bjorn Grigholm, Climate Change Institute, University of Maine.*

## OUR JOURNEY AS A CIVILIZATION

AT THE OUTSET OF THE GOLDEN AGE of climate research, we thought humans were insignificant observers, witnesses to a grand and gradual unfolding of Earth processes. Ambitious, curious, and plucky, we set out to learn what we could about how our amazing planet has evolved. What processes were under way. Its undiscovered wonders. Great adventures took place, and insights came fast. We, the authors, are awed and grateful participants in a few of these, reported here.

Along this journey, small inconsistencies became questions, which, in turn, became research projects, and then discoveries. Human activities have unwittingly set in motion worldwide tests: We have added enormous quantities of greenhouse gases, sulfate, lead, CFCs, and other compounds and toxins to the atmosphere. The resulting acid rain, diseases, loss of ozone, increased surface heating, ocean acidification, and other impacts, together with the equally significant effects of policy changes, such as the Clean Air Act and the Montreal Protocols, make belief in humanity-as-bystander unjustifable.

The past decades have revealed climate to be the synergistic sum of many parts: gases; chemicals; radiation balance; precipitation; storm and wind strength; temperature; our orbit around the Sun; variability in the energy output of the Sun; ice on land and sea; ecosystems; fires; volcanic activity; cloud cover; atmosphere and ocean circulation; and more. Earth's climate is capable of swift changes, historically unimaginable, and the results are unambiguous. Human industrial activity and our legislative decisions now overwhelmingly dominate these changes. We are not innocent, but deeply, irrevocably, involved and responsible.

The longer history of Earth's climate suggests that our activities almost certainly do not threaten the existence of life on the planet. After all, the Paleocene–Eocene Thermal Maximum around 55 million years ago was likely triggered by extreme volcanism introducing far more carbon dioxide than humanity is adding today. The effects were enormous, driving large numbers of species to extinction, deoxygenating the oceans and making them strongly acidic. Though a large portion of species perished, life survived this and other cataclysms. But recent history, and a simple look ahead, suggest that Civilization is in great danger. We risk making refugees out of hundreds of millions, if not billions, of people as water supplies disappear, precipitation patterns change, food supplies are threatened, stronger storms destroy infrastructure, and rising seas flood coastlines. We continue to expose ourselves and the ecosystem to pollutants that endanger our health and dramatically impact our quality of life.

We are collectively imposing enormous risk and cost to ourselves and future generations. The insights of this Golden Age have moved us, the authors, to expand our efforts from simple discovery to seeking to influence humanity's actions. We are, individually and collectively, responsible. The course we choose will determine the future of Civilization.

*Looking from within the penitentes on the slopes of Tupungatito, Chilean Andes. (MCM)*

## OUR OPPORTUNITY

NO MATTER WHAT WE BELIEVED DECADES AGO, there is no part of our planet that is timeless. Change is a primary characteristic of the Earth. The question is what, how, how fast, and why are the changes occurring? The long and detailed climate histories told by ice cores have revealed the nature of—and primary factors controlling—climate in the past; the dramatic effects of human activity and policy in the present; and the abrupt changes that have long characterized Earth's climate system and could revisit us as a result of our activity.

The environmental monitoring that began in Antarctica in the 1957–58 International Geophysical Year provided a clear baseline for evaluating the rise in heat-trapping industrial gases in the lower atmosphere and ozone destroying CFCs in the upper atmosphere, and identifying the climate changes caused by humans. We did not imagine, even three decades ago, that large climate changes could occur in less than a decade, much less in a year or two. Nor did we realize that our own activities could so dramatically drive climate change or mask a long-term trend—there will surely be consequences of our actions.

There is more to be learned about naturally occurring and human-caused climate changes. In order to address the threats to Civilization posed by greenhouse gases, and the serious and debilitating diseases and conditions caused by these and other pollutants, it is imperative that we ascertain their levels and distribution, and how they are changing over time. There are many more glaciers from which high-quality records of climate change can be retrieved, though they are, ironically, melting and in some cases disappearing because of warming. The exploration and scientific discovery available through these chronicles of climate change must continue until the last glacier disappears, because not having these records would mean losing the answers to questions that we may not even know we want to investigate, as was the case with the discovery of abrupt climate change. In addition to collecting climate histories, we believe that the broad publication and dissemination to the public and policymakers of real-time global and local physical and chemical climate information, along with their health and economic impacts, are critically important to defining what climate change was, is, and will be, and in stimulating an effective global response. It is essential that we all have easy access to the types of information that can severely impact our health, wealth, and security.

The past four decades have shown the significant ability we possess, but do not necessarily exercise, to reduce the risks humanity has created and the enormous opportunity we now have to provide a healthier, cleaner, safer environment, as well as a stronger economy, and a better overall quality of life for our children. For some of the pollutants we have emitted into the atmosphere, such as greenhouse gases, the danger is long-term warming and the best option we have is to reduce our emissions and plan carefully for the future. However, for industrial pollutants that do not last as long in the atmosphere, such as toxic metals, particulates, volatile organic compounds, and acid rain, the reward for reducing these emissions is reaped in days to months. For these pollutants, even in the frantic world we now inhabit, where responses and satisfaction are expected immediately, the Earth's atmosphere responds quickly enough for all of us to enjoy a healthier lifestyle, with reduced vulnerability to disease and the opportunity to see our cities and landscapes in crystal clear air now found only in remote locations, but which our ancestors enjoyed broadly, and which is within our ability to regain.

It is also time to realize that we all contributed to our current climate state and that finger-pointing only creates defensive attitudes at a time when we all need to work together—academia, industry, government, and the public—on local to global scales. We must also face the reality that not everyone will participate in the solution and that, for some, participation in the solution cannot happen immediately. But the rest of us can set the course and make sure that the "journey into climate" produces a better environment for all.

*Evening sky over a Maine shopping center. (MCM)*